VIABLE TROPICAL SMALL-SCALE AGRICULTURE

SECOND EDITION

NNAMDI CHARLES EGBUNA

Publisher
    Amazon

Printer
    Amazon

Second Edition
    November 2022

Rights
    © Nnamdi Charles Egbuna, 2008.

ISBN:
    9798370966415

All Rights Reserved. No part of this publication may be reproduced, stored in a retrieval system, cyberspace or transmitted in any form or by any means mechanical, electronic, photocopy, recording or otherwise without the prior permission of the copyright owner

Website
    nnamdiegbuna.com

TABLE OF CONTENTS

		PAGE
	Book Description	4
	Overview	5
	Introduction	7
Chapter 1	Need for Credit	16
Chapter 2	Group Credit	22
Chapter 3	Credit Institutions	28
Chapter 4	Formal Smallholder Credit	35
Chapter 5	Nigerian Smallholder Facilitators	42
Chapter 6	Smallholder Risk and Uncertainty	50
Chapter 7	Risk Management and Agricultural Insurance	60
Chapter 8	Smallholder Disaster-Risk Reduction	67
Chapter 9	Smallholder Production Planning	75
Chapter 10	Agricultural Project Failure	91
Chapter 11	Group Action	101
Chapter 12	Extension in Smallholder Agriculture	112
Chapter 13	Tropical Soil Management	125
Chapter 14	Storage in Smallholder Agriculture	139
Chapter 15	Marketing in Smallholder Agriculture	153
Chapter 16	Creating Self-Sustaining Smallholders	165
Chapter 17	Transforming Smallholder Agriculture	176
	Conclusion	195
	References	199
	Index	207

BOOK DESCRIPTION

Viable Tropical Small-Scale Agriculture is conceived as a manual, directory or reference text for stakeholders in agricultural production with special emphasis on the tropical and sub-tropical regions of the world (Africa, South-America and Asia). It also has implications for global food production.

This concise but well-researched, easy-to-understand book serves as a guide for policy-makers, consultants, educators, scholars, researchers, facilitative agencies, potential and practicing small agricultural producers. It identifies constraints to the planning, production, storage and disposal of agricultural products and proffers practical as well as innovative solutions to current and emerging problems.

There are general principles embedded therein which will serve as an invaluable guide to entities engaged in production, marketing, policy conception and implementation in non-agricultural sectors in an increasingly complex world.

The author Nnamdi Charles Egbuna holds a Bachelor of Science Degree in Agricultural-Economics from the University of Ibadan, Nigeria. He also holds a Master of Business Administration Degree in Finance from the University of Benin, Nigeria.

The author who presently works as a Private Consultant has also worked extensively in the fields of Agriculture, Audio-Visual Broadcasting as well as Information and Communications Technology. He comes from a family of writers and has three children Ngozi, Chinelo and Chinyelum.

OVERVIEW

Enhanced agricultural productivity for small-scale farmers is critical for the promotion of economic growth and poverty-alleviation as well as to avoid increased food scarcity in developing countries of the world. Well-targeted interventions by various governments have not been fully packaged to include all the components of small-scale agricultural production.

Nnamdi Charles Egbuna has packaged in this book, a comprehensive approach to small-scale farming in the tropics with the linkages between other sectors of the economy and small-scale farm productivity clearly emphasized.

This well-researched book presented in seventeen chapters provides theoretical and practical solutions to problems confronting small-scale farmers in developing countries in the tropical regions of the world. Written in a simple and easy-to-understand style, it is a down-to-earth work intended for students of agriculture, graduates of agriculture whose intentions are to become entrepreneurs in agriculture, farm managers and for policy-makers.

Agricultural production in most of the tropical developing countries has not attained its full potential due primarily to the application of inefficient technologies, poor soil management practices and massive post-harvest losses. Topics covered by the author which include; the need for credit, credit management, smallholder risk and uncertainty, production planning, extension in smallholder agriculture, the need for storage in smallholder agriculture are pertinent to the realization of increased food production.

Viable Tropical Small-Scale Agriculture is an invaluable guide for the potential or practicing small farmer whose overall objective is to go beyond subsistence agricultural production

Siaka Momoh (Professor)
College of Agricultural Management
& Rural Development
Federal University of Agriculture
Abeokuta, Nigeria

INTRODUCTION

PERTINENT FACTORS

On an unprecedented large-scale, economic, natural and human factors appear to be in a conspiracy against agricultural production in recent times, especially in the developing countries.

Increasing globalization and liberal international treaties enhance cross-border trade. This leaves communities with inefficient agricultural production, storage and disposal methods at a clear disadvantage, since they are unable to compete with cheaper imported raw and processed alternatives. The fact that farmers in the developed countries with more efficient production techniques often enjoy generous government subsidies makes the situation even more precarious for farmers in the developing countries.

Wars and communal strife which are becoming endemic in the developing countries have assumed such alarming dimensions that they now constitute significant adverse factors to the expansion and modernization of agriculture. There is fierce competition for the shrinking grazing/farm land as a result of climate change. Political, ethnic and religious differences as well as border disputes lead to violence within and between rural communities. These activities which result in the willful destruction of farm assets, disruption of farm schedules, forced migration, injury or death of members of farm-families take their toll on agricultural production.

Transnational drug cartels entice and coerce small

rural farmers into replacing the cultivation of food crops with psychotropic plants and the precursors of hard drugs. Criminal gangs engage in kidnap for ransom and brutalize small farmers into paying for protection. This makes venturing into remote farmlands both dangerous and unattractive

Local politicians who control state resources are loath to go into necessary long-term agricultural projects whose benefits may not accrue within the period of their stay in office, unless personal pecuniary gains are to be had. Frequent policy reversals also deter needed foreign investment capital.

An energy-hungry world increasingly harnesses water-power via the construction of more hydroelectric dams as sources of clean and renewable energy. This leads to periodic, alternating drought and flooding upstream and downstream. Vast areas of fertile land are thus rendered unfit or unsafe for agricultural purposes.

Exploitation of naturally-occurring resources as typified by the rapidly-growing solid minerals and upstream oil and gas extraction industries render large tracts of land unfit for agricultural uses. This situation is made worse by developed economies searching for new and alternative sources of rare-earth elements used for the production of renewable energy.

Large-scale soil, water and air pollution result from industrial accidents and other deleterious activities of corporate organizations belonging to varied industry-groups competing with illegal mining activities aimed at the extraction of precious solid minerals. These, in addition to equipment failure and vandalism often lead to cessation of agricultural activities within their immediate environs and even beyond.

Climatic changes leading to variations in agricultural production cycles and circumstances are increasingly being experienced on a global basis. Atmospheric warming fuelled by greenhouse gas emissions resulting from abuse of the environment and industrial activity is supposedly responsible for an increasing incidence of excessive rainfall, desertification, violent windstorms, altered weather patterns, flooding, drought, wildfires, heat-waves and cold spells. Salt-water intrusion raises water and soil salinity beyond the tolerance-level of many crops.

Deadly epidemics and pandemics of communicable diseases temporarily limits inter-personal communications, halts local and international travel as well as import and export activities. It also shuts down markets and transportation facilities, leaving in its wake debilitation and death of farm-family members in addition to draining scant resources and savings.

Pests and diseases of tropical agriculture infiltrate land, sea and air boundaries decimating rural farm-families and putting pressure on the resources of both the developed and tropical developing countries. Resources which otherwise would have been profitably utilized for agricultural development.

The economies of most developing countries are largely sustained by international trade. Severe fluctuations in earnings, foreign-exchange rates and domestic interest rate instability have made government-funding of agriculture errati erratic.

Pervasive corruption and mismanagement at all levels of governance exist in most of the tropical developing countries. There is also a growing need to increase defense spending to contain and deter external aggressors as well as provide subsidized energy for the restive masses. These needs restrict funding for agricultural development.

Tackling the various issues mentioned above impeding agricultural production falls outside our scope. This book deals primarily with small-scale agricultural production at the national, micro-economic and localized levels. Solutions nonetheless fall within the sphere of sincere, concrete and concerted action by the developed and developing countries to confront domestic, regional and global concerns

The entrenchment of structures that will ensure the emergence of capable leaders who understand the dynamics of international relationships and have a genuine desire to better the lives of their people through the judicious use and equitable distribution of national resources is indispensable for the mitigation of the afore-mentioned problems.

The problems addressed in this book are fundamental ones which adversely affect the typical small-scale farmer in his production environment during the normal course of production. Solutions proffered are those which can be easily adopted or adapted for use by the smallholder, policy-maker, government and other development agency in direct contact with the smallholder.

The first four short chapters of this book are devoted to issues concerning the financing of smallholder agriculture.

This is to emphasize the crucial role of credit as a highly significant factor influencing the growth of small-scale agriculture in the developing economies.

TERMINOLOGY

Certain phrases which are intermittently used in this book tend to have different meanings depending on the background and experience of the reader. These particular phrases tend to defy precise definitions. The author therefore finds it essential to explain the context in which these phrases are employed.

By deliberately avoiding the use of ambiguous terms and definitions, we will as much as possible limit the possibilities of misconceptions and controversies that may arise from their usage

DEVELOPING COUNTRIES

It is the norm to classify emerging national economies usually found in some continents such as Africa, Asia and South-America as *developing* while the heavily industrialized nations of North America and Europe for instance are known as *developed*. In Asia however, some nations which were hitherto listed among developing countries, such as South Korea, Singapore, Hong Kong (at least prior to re-unification with Mainland China) and semi-independent Taiwan are fast shedding the toga of *developing* to join Japan which is generally accepted to be a *developed* country.

In this book, our main emphasis will be on the less-industrialized economies of the tropics or hotter subtropics.

These countries possess the distinct characteristic of having low Per Capita Incomes. *Per Capita Income* is one of the indices of a country's standard of living though it is open to criticism as an objective measurement tool. Standard of living is better captured by a country's Per Capita Consumption rather than its Per Capita Income. It is also a difficult task specifying universally acceptable class boundaries (cutoff points separating the *haves* from the *have-nots*) using a cumulative-income criterion.

Another limitation arising from the use of Per Capita Income as a measurement tool is that it does not usually include the value of subsistence production; neither does it include value exchanges which occur outside the monetary system. The use of electronic-money such as Bit-coins though being an increasingly significant aspect of the financial system is yet to be fully captured and regulated.

A common denominator such as the European Euro or United States Dollar is usually chosen as the reference currency for the purpose of comparison. Value distortions often arise from the choice of exchange-rate employed for currency conversion while computing Per Capita Income. This is because exchange-rates may fluctuate rapidly as a result of random events or may be politically determined rather than being the result of the inter-play between market forces. As regional super-powers vie for influence and identity, more common denominators are emerging, further complicating the process of comparison.

Widely accepted characteristics of developing countries include; large proportion of the national population living at the poverty-line as specified by the MGI Empowerment Line which estimates the level of consumption required to fulfill eight basic needs (food, energy, housing, drinking water, sanitation, health-care, education and social security.)

The landscape of most tropical developing countries is predominantly rural with a dearth of physical infrastructure such as access roads, transportation facilities, potable water and public power supply. A large proportion of their population is engaged in agriculture and allied activities at a relatively low level of individual productivity. Many of these countries are at the earlier stages of transformation to modern exchange societies with little ties to the global economy, having just emerged from being basically closed economies not many decades ago.

SUBSISTENCE AGRICULTURE

The concept of *subsistence* in agricultural production would be misleading if taken at face value. Pure subsistence entails the production of all the food and fibre that is needed for one's survival. Obviously this situation is near impossible in today's world. Communities do exist where farmers produce most of their food requirements and other agricultural products such as natural fibres. These situations are largely sustained by isolation from external developmental influences. Some limited exchanges do exist however, even if effected mainly by barter.

There are a few near-subsistence localities in which a variety of farming activities take place in order to satisfy basic food needs. More widespread in developing countries are scenarios where producing most of the farm-families' staple food needs are combined with commercial objectives. This type of farming is called *cash cropping*, which is distinct from commercial agriculture where production is mainly for exchange purposes.

Having explained the concept of subsistence in relation to agricultural production, we will accede to the general employment of the term. That is, *subsistence* will be taken to imply agricultural production meant for personal consumption and not monetary exchange.

PEASANT FARMER

The *peasant farmer* is relatively easy to describe. He/she operates a farm-holding consisting of a few hectares of land, more or less. The farm is usually sited in a rural community where there is a dearth of such infrastructure as all-season access roads and modern communications facilities. Though labour-intensive farming methods are employed, simple
farm tools are used by a small work-force consisting primarily of family members. The peasant farmer has a low financial resource-base. Credit facilities where available are derived from informal sources. Improved farm inputs as well as advisory services are almost nonexistent or too expensive to obtain.

The typical peasant farmer has a low literacy level and little access to information on improved practices and efficient production processes. Having little control over their physical production environment, the output of peasant farmers is invariably dependent on the benevolence of soil and weather elements, pests and disease organisms. Outdated modes of agricultural production such as shifting cultivation, scattered fields and fallow periods are associated with the peasant farmer. Peasant farmers produce at near-subsistence levels leaving little surplus for exchange and/or sale in markets over which they exercise little or no control.

SMALL-SCALE

The *small-scale* farmer or *smallholder* is more difficult to describe than the peasant farmer. The major problem lies in the delineation of limits (quantities or spatial dimensions). There appears to be a lack of consensus as to the lower and upper boundaries which confine the small-scale farmer Small farmers in developing countries are usually ranked according to the size of their individual holdings. They have been variously defined as farmers whose unit of farm is limited to 5-10 hectares, 2-20 hectares or 10-12 hectares. The sole use of spatial dimensions may not give a true picture of the situation on the ground. A much better criterion for farmer classification would be the size of the *farm business*. This way, spatial dimension becomes just one of the components that determine the scale of the farmer. Size of the farm business also takes into cognizance other factors such as farm assets, labour employed, degree of mechanization, turnover, etc. It would however be best to let the reader apply the definition generally accepted in his or her locality.

Chapter One:

NEED FOR CREDIT

Evolution of Wants

In a traditional subsistence economy not tainted by external developmental influences, agricultural production is carried out basically for the purpose of satisfying the food needs of the farm-family. A little surplus food may be produced depending on the stamina of the farm-family. This surplus is conveniently exchanged for some other items which the farm-family may consider necessary but cannot produce. A farmer may consequently exchange some of his surplus produce for clothes, tools and farm-inputs, salt or some other basic need. Usually, he does not have to exchange his produce for all of his requirements because he fishes, hunts, produces and gathers some items during the seasonal periods of inactivity on his annual farming schedule.

In a subsistence economy, it appears a waste of effort producing a lot more food than is necessary to sustain the farm-family. In the closed society typical of subsistence economies, much exchange is unable to take place since the produce of neighbouring farmers is similar. The greatest hindrances to limitless food production are restricted labour availability and the level of storage technology available. Since agricultural products in their natural state tend to deteriorate within a relatively short time frame, the most that the farm-family desires is to get basic food items and seeds to last from one planting season to the next.

With the passage of time, the subsistence community

is exposed to outside influences as the closed economy gradually opens up to developmental forces. New ideas and attitudes follow in the wake of the monetization of the hitherto subsistence economy which also makes wealth easier to store.

The farm-families now want more than a thatch roof over their heads and living from one harvest to the other. Their wants now include a higher level of hygiene, a variety of exotic food items, travel, and higher levels of education, modern health-care, labour-saving gadgets, communications and entertainment devices. In summary, they now wish to attain and maintain higher standards of living and if possible, elevate their social standing within the society. These are classified as *wants* because the farm-family hitherto managed to survive without them.

The increase in scope and complexity of the farm-family's wants necessitate the acquisition of wealth-creating capabilities if their wants are to be legitimately satisfied. Where the farm-family decides not to leave farming for some other occupation, it can only obtain more income by switching to higher-income-yielding farm products or increasing the quantity of their usual products.

Smallholder Credit

In an attempt to increase the value of their farm output in order to satisfy their ever-widening scope of wants, the farm-family has to meet some additional production expenses. It has already been noted that the subsistence farmer has little tock of accumulated wealth or resources with which to meet the expenses. The expenses are often of the magnitude that cannot be easily met by his usual informal, internal and external sources of funding (savings, extended-family, friends, local moneylenders and thrift groups). It is then imperative that the farmer must seek some form of formal financial assistance. The most useful form of assistance he should obtain at this stage is Production Credit.

Agricultural (Production) Credit implies procuring money, goods or services to use immediately for agricultural production purposes while promising to pay what is obtained at a later date. Since it is most convenient to do so, agricultural credit will hereafter be confined to monetary terms.

There are compelling reasons why farmers in general need to take advantage of credit facilities in order to obtain more income from agricultural production activities; the majority of farmers in developing countries, who operate at subsistence and small-scale levels, are trapped in a vicious cycle of poverty. Their levels of production are such that they are barely able to survive from one harvest to another, leaving little or no savings. They are thus unable to convert to more modern and profitable methods of farming necessary for higher cash inflows. They obviously require credit to be able to finance the additional expenses necessary to raise output levels.

The nature of agricultural production is such that the initial activities which require considerable material resources take place quite a while before returns are made. Livestock farmers have to acquire stock and maintain them for months before they can have products for sale. The main activities in crop farming such as land-clearing, acquisition of seed-stock, planting and weeding occur long before harvest and sales. There is also the need for the deployment of increased resources within relatively short time-spans coinciding with peaks in farm activities. Credit becomes a convenient way out for farmers unable to survive the period between production and sales entirely on resources currently at their disposal.

In developing countries, local markets tend to be flooded with food items during periods coinciding with the harvest of food staples. This tends to depress the price so much so that the local farmer is not duly compensated for his/her efforts. This phenomenon largely results because of the absence of adequate preservation techniques and storage facilities. Farmers try to off-load their harvest at once in order to obtain money quickly for meeting family and other obligations such as debt repayment to local moneylenders. Credit on liberal terms will avail farmers the options of attempting storage however rudimentary and transporting produce to urban centres/population clusters, so as to obtain better prices.

Credit Effectiveness

Technology has a prominent role to play for production credit to be most beneficial to a recipient who is evolving from subsistence farming to something more substantial.

First of all, there must a better *farming technology*, ready and available for adoption. This technology would consist of improved farm tools/implements, techniques, systems of plant cultivation and animal husbandry, hardy and higher-yielding plant and animal seed-stocks, etc. This new technology must have been thoroughly tested and found to be more beneficial to the farmer than the existing one.

Any enhanced technology packaged for adoption by the credit-seeking farmer must have been demonstrated to the proposed recipient. The recipient must have also shown a genuine desire to adopt such technology if given a chance.

Concrete steps have to be taken to ensure the continuous supply of additional inputs or services required, once the enhanced technology is adopted by the farmer. These would include extension services, seed-stock, animal health-care products and agro-chemicals. These necessary inputs are to be made available at the right time, in adequate quantities and at prices which the farmers can afford.

Partial-processing options, however rudimentary, should be available to improve storage and haulage, increase the post-harvest life-span and consequently, the value of the produce. Adequate on-farm storage and infrastructure such as good road networks and water-transportation facilities should be in place to facilitate the prompt evacuation of produce at reasonable fares.

All the strenuous efforts put into adopting the enhanced technology will go to waste if there is nobody to buy the extra produce. The farmers must have an assurance that a market exists for their output without significantly depressing its unit price.

Local authorities have a major role to play in creating a conducive business environment which will attract food-processing companies and encourage them to incorporate local farmers into their production plans.

Local companies should also explore the possibility of expanding their markets domestically and exporting their products in order to mop-up excess produce by local farmers as well as encourage increased production.

Credit granted on liberal terms coupled with the above conditions would go a long way in easing the funding problems of the smallholder, which will in turn facilitate higher output levels.

Chapter Two

GROUP CREDIT

Sources

Farmers could come together for the purpose of financing agricultural projects jointly owned by them. On the other hand, they could meet for the purpose of jointly obtaining financing for their individual projects.

The sources of finance for the group's intended or current activities could be internal or external. Funds could be generated internally through the imposition of levies on group members. Periodic contributions in cash or kind are lumped together and given out turn by turn to individual members of the group or used for the group enterprise. This type of funding with its variants, date back to an era before the advent of formal banking services in many rural communities of developing countries.

There are some limitations to generating funds internally, which include the fact that only small amounts of capital can be raised. With a fast-depreciating currency in a monetized economy, the members who receive the contributions last are obviously at a disadvantage. Since there are strict timelines to be followed where rain-fed agriculture is in practice, some contributors may not receive the funds when it will be most beneficial to them. Deaths, desertions or outright inability to sustain the periodic contribution to the group fund often lead to an abandonment of the scheme at any point in time.

External sources of group funds include gifts and

grants, subvention from government and non-governmental agencies as well as loans from formal credit sources such as commercial banks. Large and multinational corporate organizations have been known to provide assistance to groups of farmers in the area of financing production activities.

Credit

Group Credit in agricultural production denotes a situation where money, goods and services are made available to farmers joined together in some form of formal or informal association, in return for a promise to repay same at a later date. Credit usually has a cost attached to it.

There are groups such as *Farmer Multipurpose Cooperative* associations whose aims are to solve a variety of their members' problems ranging from the acquisition of inputs to the marketing of their farm produce. These usually cater for the credit requirements of their members.

Thrift and Credit Union's activities revolve around the sole objective of providing finance for their members' activities. These unions vary in complexity, ranging from the informal groups to formal cooperative societies which operate like formal banking institutions. Formal unions (subject to the banking laws of individual countries) may accept savings deposit, allow occasional withdrawals and provide credit for financial members on request.

The major appeal of group credit is that groups tend to have more bargaining power than individuals. Lending agencies normally look for certain qualities in potential borrowers. These qualities include stability, ability to repay, evidence that loans would be judiciously used, security (collateral) and good record of debt payment. It is more likely that these qualities would be found within farmer-groups rather than individual farmers. Groups should also find it easier to secure acceptable guarantors or sponsors for their loan applications.

Formal credit institutions definitely prefer giving out one large loan for distribution among the farmers by themselves rather than giving out loans of relatively small amounts to individual farmers. Providing one loan for subsequent *on-lending* to individual farmers drastically reduces the cost of credit transactions. There is also added insurance against default in loan repayments since members are compelled to exert pressure on defaulters, as there is group responsibility for the loan. In the case of an individual's inability to repay, the other members have to offset the defaulter's share of the loan.

Cooperation in credit acquisition often leads to group action in some other aspects of agricultural production. For instance, inputs required for improved practices could be jointly acquired. These group activities further reduce costs which in turn translate to higher profit margins out of which the loans are repaid.

Conditions

When farmers come together for a common purpose, they are *cooperating*. Theoretically, Cooperative Associations (Cooperatives for short) are titles officially given to groups of farmers that adopt principles of operation similar to those established by the Rochdalian Cooperative Pioneers (The farmers of Rochdale County were the early pioneers of formal cooperation in agricultural production. They operated under a set of specific guidelines, many of which are still being adopted by modern-day formal cooperative societies). These principles enable them to be formally and legally addressed as Cooperatives. Cooperatives are usually registered and enjoy official recognition by relevant government agencies.

Cooperatives have a comparative advantage over unrecognized farmer-groups with regards to funding by institutionalized agricultural credit sources. In addition to government subvention, they are also able to secure funds from development institutions. Where they are insured, credit and overdraft facilities from most commercial banks are within their reach.

Group Credit could turn into a nightmare for both lenders and borrowers if a favourable environment does not exist for its implementation. First and foremost, the farmer-groups must consist of largely honest and responsible individuals. It would be an added advantage if the farmers had previously worked together to achieve set goals. Its intended uses (as is the case for credit generally) must be profitable enough to justify the attendant risks and enable easy repayment.

Sufficient fund must be made available to the group as and when required, while a ready market must exist for their products.

Constraints

One major constraint to individuals operating as a group is that it may infringe on personal liberties. The group has to institute safeguards, checks and balances in order to ensure that credit obtained is eventually repaid. Under such circumstances, everybody has to be his brother's keeper which could invade privacy and stifle individual initiative.

Some members of the group have to devote their time, energy (and perhaps financial resources) to the procurement, distribution and retrieval of the credit. Other members in some instances have been known to refuse such chores or not show enough appreciation when it has been undertaken by others. This usually results in friction or reduction of cohesion within the group and puts the cooperative venture in jeopardy.

Formal credit agencies have their guidelines and operational procedures in place. The group has to conform to these, which may be difficult or not in their best interest. Conditions to be fulfilled include; the provision of assets to serve as collateral for the loan as well as providing acceptable guarantors or sponsors.

Formal cooperative societies are not strictly capitalism-oriented groups which embrace strategies aimed at the most efficient use of scarce resources to derive maximum returns. They operate under specific principles which may not have profit maximizing as a primary objective. This may not allow for the most efficient use of credit which may be obtained at competitive interest-rates and have to be paid back.

The obvious advantages of increased accessibility to credit and the shield provided by other members in the case of personal default lends credence to the conclusion that Group Credit appears to be a worthwhile venture for small-scale farmers. Rural farmers tend to exhibit an appreciable degree of intimacy and have permanency of residence in most cases, which are essential factors that determine the success of cooperative activities. They would find it most beneficial to participate in group credit schemes since it reduces their unit cost of credit and yields quicker results than their individual efforts.

Besides, in any sphere of human endeavour, groups tend to attract and retain more attention than individuals working separately to achieve the same set of objectives.

Chapter Three

CREDIT INSTITUTIONS

Production Credit

Whatever the scale of their operations, farmers must have constant access to ready sums of money with which to finance their activities. As the size of an individual farm holding increases, it becomes more difficult for the farmer to meet all his needs solely from informal sources. Formal production credit enables a farmer to obtain necessary funding for the expansion of his holdings through his resourcefulness. Well-utilized credit facilitates the attainment of steady, self-sustaining growth.

On a domestic macroeconomic level, the ultimate goal of agricultural credit is the production of sufficient food and fibre to satisfy a country's food and industrial needs with some leftover for strategic storage purposes. Credit also boosts the production of cash-crops as well as plant and animal products destined for export to generate foreign exchange. These macroeconomic goals can be attained by stimulating the productivity of a significant proportion of the country's farmers who in the case of tropical developing countries are predominantly smallholders.

Rural dwellers that constitute the majority of subsistence and small-scale farmers in developing countries usually display a reluctance to resort to credit granted by informal sources. Constrained by a combination of pride and caution, they would only rely on credit as a last resort. There may be a rush for formal credit where it is available on easy terms, especially when it is misconstrued to be a gift or their share of the national wealth which they feel is so often denied them.

There are two distinct types of credit; for consumption purposes and for production purposes. Borrowing for *consumption* as exemplified by the case of a rural farmer who borrows some money in order to fulfill obligations to his in-laws, is quite different from a loan obtained for the purpose of increasing farm output. Borrowing exclusively for production purposes is expected to ultimately result in an increase in farm income with which the farmer pays off his debt and still has some money left over to finance future agricultural activities.

Sources

Generally, farmers can obtain credit for agricultural production from a variety of sources. These sources range from family and friends to specialized credit institutions such as commercial and development banks. These sources differ widely as each source has its own particular clientele, conditions for granting credit facilities as well as credit limits.

Agricultural production credit sources can be broadly categorized as *formal* and *informal*. Both categories have significant differences in respect of interest charges, relationship between borrower and lender, type and magnitude of collateral. Their terms may or may not appeal to the genuine farmer in search of production credit.

Informal sources of production credit include relatives and friends, shopkeepers, private moneylenders, traders, land owners, produce buyers and informal thrift and credit groups. With informal sources, procedures for securing credit vary widely while security and interest charges depend on the urgency, personality involved and the degree of risk associated with particular loan requests.

Constraints associated with informal credit are as follows; friends and relations usually have too little cash to spare and for only short time periods. The local shopkeeper expects patronage from his debtors regardless of the price and quality of his goods. The moneylender charges near-impossible interest rates, has severe penalties for default and often expects his money back in one installment. Land owners often demand exceedingly high interest returns either in cash or kind. Informal thrift and credit groups have insufficient fund to meet most of the members' needs at the same time.

Problems associated with informal credit are; severe damage to borrower's reputation and/or wellbeing in the case of default, inadequate fund to go round, excessively high interest rates coupled with short repayment periods.

Main advantages associated with informal production credit include its availability on demand if the borrower is satisfied with the conditions attached to it and the ease of access to the credit once an agreement is reached.

Formal sources of production credit include Commercial, Community, Cooperative and Development banks. These credit sources have well-defined procedures and guidelines covering tenure of loans, method of repayment and interest-rate charges. Specialized sources of agricultural finance include Islamic Banks. These institutions may not require interest on funds made available but some other arrangements such as ownership interests or profit-sharing. Other sources include Agricultural Finance Cooperatives, Micro and Rural Finance institutions.

Additional sources of formal production credit include state-run (government) or other agency (non-government) projects. Project financiers usually give out supervised credit. Such credit is given out as cash, farm inputs and services coupled with agricultural extension services. Agents also monitor the recipient farmer's activities closely in order to ensure that the credit is used prudently. It is not uncommon for the farmers to repay the loans in the form of farm-produce after harvest. The major limitation to this type of credit is that it is only available within the project locality.

Increasingly important sources of formal credit for developing countries are development projects funded by international institutions such as The International Bank for Reconstruction and Development (World Bank), Commonwealth Development Corporation (CDC), International Fund for Agricultural Development (IFAD) as well as regional development institutions such as African Development Bank (ADB). These international institutions usually work in tandem with government agencies or local banks which give out, supervise and retrieve the loans on their behalf.

While the terms for the acquisition of agricultural credit may be more lenient for formal sources, it may be harder to obtain than informal credit. The processes involved in securing formal credit can be excruciating and time-consuming. Not the least of the farmer's problems is the fact that the formal credit agencies hardly establish enough points of contact in the rural communities where the bulk of the credit-seeking farmers reside. Even though many agencies now process loan applications online via the internet, some level of physical interaction is still required.

Formal lenders usually ask for high levels of collateral (security). Where the collateral requirements are softened, the agencies are so inundated by loans requests that they are unable to meet a significant proportion of the requests. Another consequence of softened collateral requirements is that corrupt loan officials in some developing countries have been known to constitute themselves into *tollgates* to be passed through if loan applications are to be successful.

Formal applications for production credit involve the provision of quantitative as well as qualitative data necessary for the evaluation of the applicant's request. This is applicable even for funds provided through Interventions targeted at entities adversely affected by such unpredictable phenomena as pandemics and natural disasters. The illiterate or barely literate smallholder may be unable to perform the mathematical computations and predictions which are expected to have been deduced from farm records which up to that stage are nonexistent.

Choice

There are some important preliminary investigations which a farm operator ought to carry out in order to determine his choice of credit source and the amount of credit he can conveniently manage. It is here assumed that the smallholder only has access to seasonal or short-term credit facilities. This derives from the notion that credits agencies operating in tropical developing countries are usually in a hurry to cut their losses retrieve and recycle their capital so as to increase profit, reduce the level of risk or satisfy other applicants waiting in line.

Any farmer in need of credit should have a foreknowledge of the likely impact of credit on his farming operations. In other words, he has to determine his probable levels of production with and without the use of credit. He also has to attempt a near-accurate forecast of cost of inputs for the production process and product prices at harvest time. These predictions are not easy for the smallholder faced with the uncertainties of weather, personal concerns and unfavourable government policies. He is not even sure of the extent to which bumper harvests or general increased output would negatively affect output prices in the absence of adequate processing, storage and preservation facilities. Nonetheless the farmer has to do these necessary computations. Experience comes into play at this stage since an older farmer would probably be more adept at forecasting than a newer farmer

Resort to professional help by competent persons where available and affordable, may be in order. The professional is able to use sensitivity analyses to create possible production scenarios for the farmer, while watching out for macro-economic issues on the horizon which may affect the production environment.

Factors that will make a farm operator choose a particular formal credit source over another include cost of the credit. Cost of credit encompasses interest rates as well as expenses and all the formal and informal payments made in respect of the credit applied for. The farmer also takes a critical look at the penalties imposed by each source on loan defaulters. He would normally prefer not to lose his tangible farm assets, given a choice. The smallholder may neither have the resources nor inclination to make repeated journeys over long distances in order to transact business with a credit institution. Timeliness of loan disbursement is another critical factor that would determine the preferred loan source.

The choice of a preferred credit source by any farmer involves a balancing of the favourable and unfavourable conditions and requirements of each available source. These are considered in the light of his peculiar production circumstances in order to make an informed decision. The quest for a tailor-made source suitable for adoption by all smallholders would be a fruitless venture since each source has its obvious merits and limitations.

Chapter Four:

FORMAL SMALLHOLDER CREDIT

Commercial Banks

Commercial banks constitute an ideal vehicle for the distribution of small-scale agricultural production credit. In addition to being the most common category of credit institutions based in multiple locations to be found in many developing countries, they have the added advantage of having a large cumulative resource base. Regrettably, they do not appear to willingly subscribe to the notion that the smallholder who produces the bulk of the food and fibre in most developing countries should have a significant place in their credit operation strategies. They often give near-impossible conditions to small farmers in search of credit.

First and foremost, commercial banks are in business to make acceptable profit margins for their shareholders and therefore wish to obtain the highest possible returns on funds in their loan portfolio. Conditions inherent in the provision of agricultural credit may not be in consonance with profit maximizing objectives. Commercial banks do not consider agricultural credit as a preferred investment option because of its significant levels of uncertainties and risks, peculiar requirements and high operational costs.

Commercial banks may not be inherently structured to give out agricultural credit to energize small-scale production if they are plagued by such operational weaknesses as ill-conceived credit policies and poor fund management abilities. Other possible problems they may encounter include; lack of, or inadequate incentives to give out smallholder loans, periodic non-availability of funds, interest-rate ceilings and high incidence of non-performing loans.

On a cursory examination, agricultural credit appears to be a complex choice within a portfolio of investment options. There are ever-present possibilities of largely unpredictable and unfavourable weather conditions, disease and pest infestations as well as possible farmer incapacitation. The amount of money required by the smallholder is usually small in relation to processing costs. It is not uncommon for a smallholder's farm plots to be situated in different locations making monitoring costly and time-consuming. The existing level of technology is low coupled with a profound lack of necessary infrastructure such as storage facilities and all-season access roads to enable the rapid evacuation of farm produce. Agricultural production in its natural state has some peculiarities such as wide seasonal fluctuations in output and consequently, prices. The literacy-level among smallholders is generally low so most of them are unable to keep comprehensive records of their farm operations. Dishonest individuals have been known to masquerade as farmers in order to obtain and divert loans specifically given out as agricultural credit. The biological nature of agricultural production makes computations difficult since there are no universally accepted formulae for the valuation of plants and animals at various stages of growth. The scheduling, inspection and monitoring of agricultural credit require staff with specialized knowledge, which the banks do not have in sufficient numbers. In cases of default, collateral (which are more often than not, landed property and/or other tangible assets) are not easy to dispose of since they are usually sited in rural localities.

The involvement of commercial banks in small-scale agriculture may however prove counterproductive in situations where profit-maximizing is their sole objective. In such situations, savings mobilized from small farmers situated in rural areas are usually deployed in urban areas and to non-agricultural projects which are considered less risky with shorter repayment periods. This could result in further de-capitalization of the rural areas.

Credit Viability

There appears to be quite an impressive muster of reasons to justify the commercial bank's reluctance to provide smallholder agricultural credit when viewed from the bank's corporate perspective. Commercial banks are however an integral part of the larger economy which unequivocally agrees that there is a compelling need to break the vicious cycle of low lending - poor inputs - low yields - poor repayment which is the bane of the small-scale agricultural sub-sector. There are enough reasons to justify their active participation in the provision of agricultural credit. If well-conceived and correctly administered, agricultural credit can be provided by commercial banks without necessarily incurring unacceptable losses.

Rural infrastructure is gradually improving in many developing countries which enjoy relative peace and economic stability. Commercial banks could leverage on technology to fine-tune on-line platforms for managing loan applications. This will drastically reduce the magnitude of processing costs associated with agricultural credit.

Beside goodwill from identifying with rural projects and social responsibility obligations, banking business thrives best within a robust and growing economy. Facilitating agricultural production which contributes to national economic growth is ultimately to the advantage of the banking sector.

Agricultural credit can take its rightful place among the portfolio of viable investment options if a number of measures are taken. Most importantly, qualified personnel should be employed in sufficient numbers to assess credit proposals and supervise ongoing projects. As an alternative, commercial banks could have periodic in-service workshops and training for regular staff, aimed at improving their agricultural-credit management skills.

Crop production under rain-fed agriculture requires precise timing of the necessary tasks. Leveraging on the increasing availability and precision of weather-forecasting agencies, commercial banks should make credit facilities available as and when needed. Releasing funds too early or too late increases its likelihood of misuse or diversion. In addition, draw-down on approved loans should be extended to two or more installments in tune with peaks in resource-needs within the production cycle. Direct payments should also be made by the banks to input-suppliers or service-providers where practicable.

The review of viability of ongoing projects should be on a continuous basis to minimize losses due to unforeseen circumstances. However, where projects are in danger of being abandoned due to unfavourable policies such as currency depreciation, natural and man-made disasters or other adverse situations, banks should be ready to refinance or increase the level of funding in order to rescue the project. It may also be necessary to extend the time-frame for loan repayments.

Decision-making processes for agricultural credit should be decentralized. Competent rural bank managers could be given the authority to give out predetermined levels of credit. They should also be permitted to monitor and evaluate projects thereby reducing operational costs. Their activities must however be closely monitored to avoid the abuse of this authority.

Formal agricultural credit institutions should establish a common information-pool on prospective and operating credit beneficiaries. Some countries have centralized bank-account holder identification systems like Nigeria's Bank Verification Number (BVN). These measures will help to prevent multiple credits for the same projects, which invariably lead to defaults in loan repayment. These institutions should also sponsor enlightenment programs aimed at minimizing smallholder loan default.

Employing *On-Lending* strategies, commercial banks should encourage smallholders to apply for credit in groups. A lump sum could be advanced to the group for subsequent distribution among members. Further loans should be made available on easier terms to small-scale farmers who have proved their creditworthiness by making payments on time and in full.

Government Roles

Governments of developing countries usually provide or facilitate the provision of physical infrastructure such as road networks, communications facilities, power, research, mechanization, market and processing facilities. These infrastructures pave the way for a smooth credit delivery, monitoring and retrieval system by reducing communication, production, transportation and marketing costs.

Facilitating credit for the agricultural sector boosts its *Value Chain* made up of its production, processing, storage, farm-input supply and marketing aspects. This ensures Food Security and adds to the national Gross Domestic Product.

Lending to the service sector is less cumbersome and takes precedence over lending to the real sector as far as commercial banking is concerned. Governments, as a macroeconomic development strategy should enact policies and institute schemes aimed at encouraging commercial banks to give out agricultural credit. These schemes should be primarily designed to shield the commercial banks from excessive financial losses in the event of a persistent inability of smallholders to repay loans granted to them.

Governments should enlighten the smallholder on the advantages of production credit, loan discipline, wise use of credit and the necessity of loan repayment when due and in full. Enlightenment can be carried out at the various points of contact between the smallholders and government agencies, by the use of extension workers and through government-controlled mass-media organs.

It is hoped that a dissection of the reasons for the reluctance of commercial banks to grant liberal smallholder agricultural credit as well as the roles expected of banks and other concerned agencies, will pave the way for an increased volume of credit transactions to the benefit of the economies of developing countries.

Chapter Five

NIGERIAN SMALLHOLDER FACILITATORS

Development Agencies

A cursory look at the development agencies operating within the Nigerian small-scale agricultural sub-sector will enable policy-makers to conceptualize, organize or fine-tune structures that will bring about a rapid transformation of their moribund smallholder agricultural sectors.

The choice of this particular country's agricultural sector does not imply that it has achieved resounding success at smallholder agricultural transformation. There is more to achieving success than just having enabling structures and sound policies in place. Capable personnel given adequate resources with the right motives facilitate the process.

Creating the right environment in which smallholder agriculture can thrive is the first indispensable step towards its transformation and development.

Structure

There are three tiers of governance in the Nigerian State consisting of Federal, State and Local governments. These tiers are expected to actualize the broad *National Agricultural Sector Objectives* which include Food Security, Provision of Fibre as well as Industrial-sector Raw Materials, Employment Creation, Foreign-Exchange generation, Rural Development and Stimulating Demand for other industrial products.

At the Federal level, the national agricultural policy formulation is carried out. Strategies and capacity for carrying out these policies are also put in place. Agencies such as Federal Ministry of Finance and Federal Ministry of Agriculture and Rural Development (FMARD) have its' activities impacting on agriculture. The Federal Capital Territory is the only land area directly administered by the Federal tier. It is therefore inevitable that the implementation of agricultural policies largely devolves to the states and local governments.

Each state has a government ministry solely devoted to agriculture or a department of agriculture in some other ministry. These ministries/departments are pre-occupied with such issues as Credit Schemes, Extension Services, Projects Coordination, Development Projects, Farm Structure, Inorganic Fertilizer Procurement and Distribution, Integrated Rural Agricultural Development, Farm Erosion Control, Produce Inspection, Partnership Projects, Training, Livestock Development, Farm-family Welfare Programs, etc.

The Local Government statutorily has a role to play in agricultural development. An overwhelming majority of the 774 local governments in Nigeria are rural-based and are largely dependent on agriculture. Theoretically, they are in a vantage position to ensure the active participation of traditional institutions and peoples in schemes initiated and directed by the Federal and State Governments. They should also provide home-grown initiatives to respond to local needs and conditions. Sizeable projects undertaken by State or Federal Governments in partnership with international agencies are usually actualized at the local government and community levels.

Joint Federal / State Projects

There are collaborative efforts between the Federal Government and the various state governments to develop the agricultural sector. First of all, the Federal Government serves as a guarantor of all foreign loans secured by the individual states. It is also responsible for the payment of some Counterpart Funding required by foreign entities collaborating with the Federal and State governments to execute specific projects within Nigerian states.

Agricultural Development Projects (ADPs) which are joint Federal/State entities exist with the stated objectives of making agricultural inputs and services available as needed and within the easy reach of farmers. They are conceived as decentralized farming and input distribution centres.

River Basin Development Authorities (RBDAs) established by the Federal Government for the states have the primary mandate to develop water resources for use in irrigated agriculture (small dams, wells, bore-holes, drainage systems, flood and erosion control) It has since expanded its' mandate to include general agriculture and the development of other physical infrastructure. They are now involved in the provision of land clearing services, feeder roads, rural power and agro-service centres.

Central Bank of Nigeria

The Central Bank of Nigeria (*CBN*) is at the apex of the Nigerian banking sector and in conjunction with the Federal Ministry of Finance constitutes the twin pivots around which the country's financial activities revolve. CBN is used by the Federal Government of Nigeria (FGN) to conceptualize and implement monetary policies aimed at the smooth operation of the national economy.

The CBN oversees the activities of Nigerian banking institutions which consist of Commercial banks, Micro-finance banks and Development Finance institutions. The shareholders of commercial and micro-finance banks comprise of individuals, corporate entities, cooperatives, communities, state or local government authorities. Development banks and Finance institutions are however solely government owned.

The CBN in addition to its core responsibilities plays a role in enlightenment and education by collecting, collating and publishing information which are of importance to agriculture and other sectors of the economy.

In collaboration with such international development institutions as the International Bank for Reconstruction and Development (World Bank), Commonwealth Development Corporation (CDC), International Finance Corporation (IFC), African Development Bank (ADB) and International Fund for Agricultural Development (IFAD), the CBN executes projects in such areas as agricultural production, irrigation and livestock development. .

Micro-finance banks are essentially regulated limited finance institutions. Banks which were hitherto referred to as Community Banks were directed to convert to micro-finance banks. According to CBN, *micro-finance* consists of smaller loans, deposits and financial services to the poor who lack access to other formal financial institutions. The rationale for the establishment of micro-finance banks revolves around the need to increase the share of micro finance (micro-credit) to the national economy as well as eliminate gender disparity (discrimination) in access to formal financial services. Micro-credit is thus expected to serve as a catalyst for poverty reduction and boosting economic growth.

The Agricultural Credit Guarantee Scheme Fund (*ACGSF*) jointly owned by the FGN and CBN and supervised by the CBN guarantees commercial bank up to seventy-five percent of the loan in default for agricultural purposes. Innovations of the scheme include Self-Help Group Banking, Trust Fund Model, Intermediation and Interest Draw-Back.

Self-Help Linkage Banking denotes a variant where groups of farmers can obtain loans in multiples of the balance in their savings account at the time of loan application. In the Trust Fund Model, states, local governments and other agencies place funds to augment the small group savings of the farmers as security for agricultural loans. *Agricultural Finance Intermediation* involves drawing up a Memorandum of Understanding involving an interested party (for instance state-government, private foundation, multinational or non-governmental organization), a lending bank and the CBN, specifying duties and bligations of each party in respect of the loan under consideration. The Interest Drawback Programme provides for interest rebate where loans are repaid as and when due.

Specialized Institutions

Some specialized institutions exist whose mandate

enhances smallholder agricultural production. The CBN and the Federal Ministry of Agriculture and Rural Development (FMARD) are responsible for establishment and supervision of some of these institutions which include the Nigerian Agricultural Insurance Company (NAIC), Nigerian Agricultural Cooperative and Rural Development Bank (NACRDB), Bank of Agriculture (BOA), Nigerian Agricultural Quarantine Service (NAQS), Nigeria Incentive-Based Risk-Sharing System for Agricultural Lending (NIRSAL) and National Programme for Agriculture and Food Security (NPFS).

NAIC is wholly owned by the Federal Government and induce the provision of credit to farmers, provide emergency assistance, promote production, provide financial support for losses from natural hazards. Encouraging farmers to adopt high-value inputs and improved farming practices are also some of its major objectives. NAIC addresses the reluctance of the conventional insurers to carry agricultural risks by administering the agricultural insurance scheme which subsidizes premium payments on selected crop and livestock policies. Crops are insured against unexpected loss of projected yield and/or loss of projected profit. Farmers are consequently granted financial protection and loss reduction from accidents, natural perils and calamities such as fire, lightning, drought, flood, pests and disease outbreaks.

NACRDB is a development finance institution under the direct supervision of the Ministry of Agriculture and Rural Development. NACRDB extends loans to state governments, corporate entities, cooperatives and individual farmers. It is involved in on-lending, smallholder schemes and livestock development programmes. It also provides guarantees and makes direct investments in the equity capital of major agricultural and agro-allied ventures.

BOA is a development finance institution wholly owned by the Federal Government providing credit facilities to support agricultural value-chain activities (production, processing, storage, farm-input supply and marketing). Its other functions include non-agricultural micro-credit, savings mobilization, technical support and financial services advisory.

NAQS was set up basically to prevent the introduction and establishment of animal and zoonotic diseases as well as pests of plants and fisheries and their products. Their activities also cover import and export products and items.

NIRSAL established in conjunction with Alliance for Green Revolution in Africa, uses the strategy of employing the tools of Risk-Sharing, Agricultural Bank Rating, Insurance, Technical Assistance and Bank Incentives in an integrated approach aimed at breaking the stagnation and decline in the agricultural sector.

NPFS coordinates donor-assisted agricultural projects. It also cooperates with relevant departments of agriculture to harmonize activities aimed at boosting production.

Interventions

Government on its own or in partnership with other local agencies such as NIRSAL or international agencies such as World Bank or IFAD may choose to embark on limited, time-bound interventions to assist households. These interventions may be in the form of targeted credit facilities or grants to ameliorate the effects of such phenomena as pandemics, insecurity, natural disasters such as wildfires and drought.

Chapter Six:

SMALLHOLDER RISK AND UNCERTAINTY

Agricultural Production

There are risks and uncertainties facing every course of action because its consequences occur in the future.

A course of action is said to be at risk when its possible consequences or outcomes are known and probabilities can be assigned to them. That is, the magnitude or frequency of occurrence of each possible outcome is known or can be estimated. In the case of an uncertainty, the possible consequences of a particular course of action are known but the magnitude or frequency of their occurrence are not known and cannot be estimated.

Farmers are constantly confronted with situations of risk and uncertainty since their present decisions have future consequences. Decisions taken in the course of agricultural production have their appropriateness reflected in the magnitude of profits or losses recorded at harvest time. Generally, the levels of risk and uncertainty associated with agriculture are significant. These high levels are as a result of some characteristics of agricultural production such as the long production cycle. The time interval between the initiation and completion of agricultural production is relatively long in comparison to typical industrial production processes. This is usual with projects involving tree crops and large farm animals. It is possible that favourable conditions which prompted their initiation may no longer exist by the time the crops or animals are ready for harvest several years later.

One notable peculiarity of agricultural production is the *weather factor*. Though certain agricultural production processes such as intensive livestock farming can be carried out in a fairly-regulated environment, most smallholder productions are subject to the dictates of climatic conditions. Weather extremes such as heat-waves and windstorms, irregular, insufficient or excessive rainfall pose constraints to normal agricultural production processes.

There is a *competition factor* with regards to agricultural production. This denotes a situation where man in his quest for food and fibre competes with other living organisms for sustenance and space. Weeds compete with domesticated plants for space, water, nutrients and sunlight. Birds, rodents and insects frequently attack agricultural produce in their pre and post-harvest stages. Microscopic organisms cause deterioration and debilitation in plants and animals.

Another notable feature of agriculture is its vulnerability to natural and man-made disasters. These disasters could occur in such forms as floods, drought, soil contamination or destabilization, wild-fires, soil erosion, pest infestations or disease outbreaks.

In the developing countries, unceasing efforts are being made to initiate and sustain rapid agricultural sector growth. Farmers expand output, adopt new species or varieties, employ new techniques or venture into new areas of production. These efforts further increase the already significant levels of risk and uncertainty.

ENVIRONMENT

The agricultural production environment in Nigeria typifies the types and magnitude of risks and uncertainties faced by the smallholder in developing countries. A large proportion of the Nigerian population lives in rural communities. The predominant occupation of these communities is small-scale agriculture which is carried out at near-subsistence levels. These farmers derive little income from farming activities, which after the deduction of expenses and other obligations leave them with little or no savings. This low income situation exists side by side with high birthrate, lack of social welfare amenities and extended family obligations. They are in most cases, unable to finance the expansion of their farm activities out of their personal resources. Because of the prevailing low levels of literacy and awareness, lack of adequate collateral and apathy by formal lending agencies, they are unable to secure production credit on favourable terms.

The rural smallholder tropical farmers usually do not have adequate technical support, all-season access roads, adequate transportation facilities, potable water and modern health-care. With the existing level of technology, production depends entirely on the benevolence of weather elements. There is almost total dependence on rainfall. Where irrigation facilities are available, lack of fuels, oils, pumps, hoses, other accessories and consumable items prevent their full and effective utilization.

The labour-intensive techniques of production are carried out with rudimentary implements, on a much-abused soil. In the absence of mechanical and chemical pest control facilities, the farmers have to contend with rodents, birds and insects for the products of their low-yield seed-stock. In extreme cases, some pestilence or natural disaster could wipe out their entire investment before it even gets to the harvest stage. Where the farmers are able to harvest a significant proportion of their produce, there are inadequate storage facilities. They consequently have to find the means of transporting it to local markets already filled with similar products from nearby farms. In this situation of market saturation, they have to find buyers for the produce before it deteriorates further and become unfit for human consumption, industrial use or other commercial purposes.

Differentiation
The distinction between risk and uncertainty may appear inconsequential to the smallholder who is preoccupied with trying his possible best in the present and sees no alternative but to consign future outcomes to providence. A clearer understanding of risk and uncertainty in smallholder agricultural production will however enable farmers and other pertinent parties (policy-makers as well as facilitating and collaborating agencies) to have a better chance of achieving their micro and macroeconomic objectives in relation to agriculture.

Risk situations can be managed, since their magnitude and occurrences can be determined. *Risk management* in agricultural production involves the identification, measurement and control of consequences arising from decisions which threaten the income and assets of the farmer. A smallholder who has been cultivating a particular specie of crop for a number of years and decides to plant an additional hectare of the same species is able to anticipate the types and severity of the problems as well as what it would take to solve the problems. An individual who is contemplating going into agricultural production as a business may also be facing a risk situation where he/she has acquired a working knowledge of the requisite production processes.

In risk situations, farmers are able to plan their activities in such a way that minimizes the effects of specific unfavourable outcomes. The main risk faced by farmers is the possibility of fluctuations in output and consequently income. Fluctuations in output could result from droughts, floods, weather variations, fire outbreaks, pestilence and diseases. Other risk factors are theft, labour apathy, death or incapacitation of the farm operator. The risk factors enumerated above are not all-encompassing.

Uncertainty in smallholder production is exemplified by a small-scale poultry farmer in a developing country who decides to pioneer wheat production within his locality in order to increase his income and better his lifestyle. This farmer faces a situation of uncertainty because he has neither prior experience of wheat-farming nor a neighbour from whom he can learn the fundamentals of wheat production, even if the prevailing weather conditions are suitable for its cultivation.

Situations of uncertainty for a smallholder include changes in cost of inputs and product prices. This *Price*

Uncertainty exists because of the diverse factors involved in the mechanism of price-determination. It is not possible to accurately predict the magnitude of change and in what direction costs and prices would shift in the future, beyond a specific time-frame which is usually of a short duration.

Yield Uncertainty exists where the peculiarities of domesticated plants and animals are not well known or their techniques of production not yet mastered.

Innovation and new research findings may render an operational system of production inefficient. Unit costs under the old system may be excessive in relation to the prevailing market price if the adoption of the improved system by other producers have become widespread. This possibility is known as *Technological Uncertainty*.

A situation of uncertainty could arise from *institutional* sources. Government policies such as frequent currency devaluations could affect real disposable income, thereby affecting agricultural consumption patterns.

Uncertainty could also arise from *legislation* directly or indirectly affecting agricultural production. Instances include bans on the export or the unrestricted import of some staple local products. Restrictions may suddenly be placed or taxes levied on some derivative products such as cigarettes made from tobacco leaves. Government could decide to enforce laws prohibiting juvenile labour. This would have the effect of depleting the farm-family labour force.

Planning

The subsistence or small-scale farmer in the

developing country is largely misunderstood. He is stereotyped as an unwilling recipient of innovation. Policy makers are not willing to take into cognizance past defective policies and the unfriendly production environment which have taught the smallholder to act with extreme caution. Farmers should not be coerced into abandoning their seeming cautious behaviour. A more productive way of re-orientation is to make the farmers realize that there is much prudence in thorough production planning. Advantages to be had in planning for risk and uncertainty derive from the fact that mishaps during the course of production will have as little adverse effect as possible on anticipated yields and incomes.

Generally, *planning for risk and uncertainty* in agricultural production starts at the initial stages before actual production commences. Decisions made at the conceptual stage will either minimize or aggravate the fluctuations in future yields and incomes. In planning for risk and uncertainty, there are certain options open to any farmer. These options include maintaining monetary reserves, adequate knowledge of proposed project, building contingency allowances into farm budgets, employing experienced hands, taking advantage of price guarantees and agricultural insurance. Other options include product specialization in order to concentrate efforts and enjoy economies of scale, flexibility (employing multi-purpose plants and equipment), mixed cropping and mixed farming.

Options

Reverting to the scenario painted of the typical

smallholder in the developing countries, it is obvious that most of the planning options enumerated above are beyond his reach. The peasant farmer who has barely enough resources to commence seasonal production has no monetary reserves to hold. He probably does not know what a farm budget looks like; neither can he afford professional help. Since his level of production is low, he cannot contract prices and costs directly but has to go through middlemen who incorporate their own profit margins. Agricultural Insurance is a recent phenomenon in the developing countries and cannot be heavily relied upon. Moreover, agricultural insurance usually covers selected crops and livestock and as such is not accessible to many smallholders.

The smallholder's limited options in planning for risk and uncertainty does not infer that he is entirely defenseless against adverse production and other circumstances. Even an illiterate farmer can take out personal insurance against accidents and physical incapacitation if he can meet the required premium payment obligations.

In planning for risk and uncertainty, the first option open to the smallholder is to stick to enterprises which he knows are reliable by personal experience or wide consultation. The farmer should choose between and within animal and plant species and decide to embark on the production of those which are most suited to his peculiar circumstances.

The smallholder could diversify his production activities. The rationale for this is that the probability of a series of unrelated enterprises failing at the same time is considerably lower than for related enterprises. Diversification spreads out risks and uncertainties encountered during the varied production processes. *Diversification* could occur in such forms as different farm locations, multiple-cropping of unrelated species, cropping of the same species at different times (especially in irrigated agriculture), crop rotation and mixed farming involving combinations of plants and animals (for instance intensive poultry/maize where the crops can be protected from the animals)

The smallholder could maintain flexibility of production in respect of time and equipment to some extent. Flexibility facilitates a quick change from one enterprise to another since no rigid production plan is followed. For instance, the management of laying hens could be carried out in deep-litter instead of in battery cages to enable the quick conversion of poultry sheds to other uses, should the necessity arise.

The smallholder could, in partnership with neighbouring farmers, embark on contract-farming. Contract-farming reduces the risks and uncertainties emanating from changes in cost, price and demand patterns. Contract-farming promotes a degree of income stability and assurance since fluctuations in market prices would not affect the farmer's income expectations.

Some miscellaneous strategies which may form the core of planning for risk and uncertainty include; fencing farm holdings to deter encroachment to some extent, improving sanitation by constant weeding and the clearing of farm borders to discourage pests, using disease-resistant hybrid or genetically modified seed-stock, monitoring of endemic pests and diseases as well as deferring planting to the last possible time period. Group action on security and other pertinent matters can also be explored.

In these days of fluctuating weather patterns, species with earlier maturation periods are preferable, if available. Farming in the usually fertile but low-lying flood plains should be approached with greater caution. Increased emphasis should be placed on the use of data from short and long-term weather forecasting stations, in planning for production.

It is essential that farmers should understand that there is a trade-off between the quest for profit maximizing and planning for risk and uncertainty. This trade-off occurs because risk planning has a cost attached to it in the form of foregone income. For instance, mixed-cropping leads to reduced output per hectare of farmland. This is as a result of competition between the different species, which in turn results in lower farm income. Similarly, the contract-farmer forfeits the chance of windfall gains in the event of increased market prices at harvest time. Generally, the more profitable agricultural projects are associated with higher levels of risk and uncertainty.

All said and done, it is definitely better for a smallholder to forgo a small part of his anticipated income in order to ensure that he receives a substantial part of it, by exercising due caution.

Chapter Seven

RISK MANAGEMENT AND AGRICULTURAL INSURANCE

The Concept of Risk

Agricultural Insurance, simply put, involves the transfer of responsibility for farm losses from the farm operator to another entity for a fee.

Risks common to agriculture can be conveniently categorized into four groups for the purpose of insurance. These are Natural, Biological, Economic and Social risks. Natural risks include drought, excessive rainfall and flooding, natural fire outbreaks, windstorms and erosion. Biological risks involve such phenomena as pest upsurges and disease incidence. Economic risks are basically concerned with changes in market forces leading to adverse fluctuations in anticipated income while Social risks have to do with theft, willful destruction and related issues.

Fundamental problems facing a farm operator are; how to protect his assets, ensure his expected income and stay in continuous and profitable production. These problems lead the farmer to the threshold of Risk Management. Risk Management involves the analysis of risk situations for the purpose of evolving feasible control measures.

Risk Management entails making adequate arrangement for possible losses, taking into consideration available resources and resources required to ensure a successful execution of a project. In agricultural production, this would imply the successive stages of risk identification, analysis, control, finance and monitoring.

Risk Identification requires a thorough knowledge of the production environment and process. Armed with this knowledge, a comprehensive list of possible consequences of a particular course of action can be compiled.

Risk Analysis involves the estimation of the frequency and magnitude of each occurrence previously noted during the identification stage. (In situations of uncertainty, such frequencies and magnitude cannot be estimated.) During analysis, the magnitude of losses sustained in the event of unfavourable outcomes is determined. The outcome of Risk Analysis largely depends on the degree of exposure of the project, current state of the project and up-to-date records.

Risk Control measures include Risk Assumption/Reduction, Risk Prevention, Risk Combination and Risk Transfer.

Agricultural Insurance is a form of Risk Transfer which enables the farmer to convert the possibility of a large loss of income into a relatively small predetermined cost known as *premium*. This has the effect of reducing the impact of adverse production circumstances thereby ensuring the stability of farm income. Besides farm-income stability, agricultural insurance can be used to achieve some other objectives such as; articulating better farm management strategies, adopting new and improved farming practices, facilitating greater investment in agriculture, ensuring farmer stability in the wake of disasters and facilitating access to formal credit.

Forms

Agricultural Insurance policies come in different forms, the most familiar being Multiple Crop Insurance, Single Animal/Crop Insurance, Specific Risk Insurance, Combined Risk Insurance and All Risk Insurance. Fundamental to the success of an agricultural insurance scheme are sufficient funding and the existence of capable machinery for its implementation.

It should be noted that agricultural insurance does not operate in isolation from other schemes necessary for the success of agricultural projects. It is certainly not a panacea for all problems faced during the course of production.

Constraints

There are constraints which could impair the successful execution of agricultural insurance schemes in the developing countries. These include inadequate or nonexistent storage facilities, low level of technology and poor production techniques, poor communication and transportation facilities, lack of social and health-care amenities and low product-prices.

The complexities of Project Valuation and Loss Determination pose formidable constraints to the success of agricultural insurance schemes. The agricultural sector of a typical developing country is dominated by smallholders who are predominantly engaged in mixed farming. Cost of damage to crops is difficult, if not outright impossible to compute in mixed systems. A related issue is the lack of relevant and eliable data on previous losses as well as premium payments.

These phenomena make the calculation of equitable premiums and compensations very difficult tasks. In addition, some elements of agricultural risks cannot be conveniently insured against as a result of excessive administrative costs. Grain losses in storage and egg breakages serve as examples.

Worthy of mention is the issue of timeliness in the settlement of premiums and indemnity. *Premium* constitutes periodic payments made to the insurer in exchange for accepting to shoulder the farmer's possible losses. *Indemnity* is the compensation made by the insurer in the event of insured losses by the farmer. Considering the circumstances of the smallholder in a developing country, it is not certain that most of them would be able to make timely premium payments, which are essential for the success of any insurance scheme. It may also take a while for the insurer to verify the authenticity of genuine claims before paying compensation. Time which a seasonal farmer can ill-afford.

Governments with their cumbersome bureaucratic structures often have to take the lead in conceiving, sponsoring, financing and often times implementing small-farmer insurance.

Another constraint to agricultural insurance is the role of *intermediation* in the industry. The insurance business operates through a system of brokerage or commission-men. The success of an agricultural insurance scheme largely depends on the degree of commitment and margins earned by these middlemen strategically located between the farmer and the insurer.

Other constraints to agricultural insurance in developing countries worthy of mention include low insurer resource-base, low level of farmer-awareness, apathy and caution in dealing with government-sponsored programs especially where money is to be paid out, limited access to schemes which are implemented in partnership with formal credit sources, small-size and fragmentation of farm holdings, faulty or discriminatory land allocation and cumbersome land-tenure systems, lack of farm records in addition to insurer personnel and other inadequacies.

Strategies

The magnitude of problems facing implementation of agricultural insurance in developing countries, require the adoption of peculiar strategies in order to ensure its success.

At the conceptual stage, there should be conscious efforts to sell the scheme to its target recipients. Insurers should provide informational materials at prime locations such as council headquarters, input-supply stores and local markets. Insurers could also take advantage of local seminars and training workshops held for small farmers. As many brokers as possible should be involved in order to enhance accessibility. Commissions earned by brokers should be kept in check while safeguards should be put in place to prevent collusion to short-change farmers..

Schemes which appear to satisfy many of the requirements for success, including adequate funding and facilities have been known to falter as a result of poor implementation. The management cadre for agricultural insurance schemes should consist of people with proven competence and a working knowledge of production processes. Field personnel should possess such qualities as dedication to duty and patience.

There is need for continuous publicity campaigns that will heighten the small-farmer's awareness of the existence and advantages of the scheme. The campaigns should ideally be available to farmers when they are not busy with farm activities.

Generally, insurance companies in developing countries have acquired notoriety for prompt collection of premiums and marked reluctance in settling genuine claims. Agricultural insurance schemes must ensure that they deviate from this apparent norm in order to obtain and retain the confidence of client-farmers and participating formal credit agencies.

Governments should introduce measures that would deliberately integrate insurance into routine agricultural production processes. For instance, the possession of insurance certificates could serve as a prerequisite for obtaining subsidized farm inputs. Policy measures should aim at getting more formal agricultural-support and facilitating organizations involved in insurance schemes.

More frequent and intense weather variations are being experienced on a global basis. Disaster Risk Reduction strategies are consequently assuming more prominent roles, not only in small-scale agriculture in the developing countries, but for rural development as well. Funding for Disaster Risk Management should gradually become a very significant component of national, state and local budgets since it is a social cost to be borne in order to ensure balanced economic development create employment and enhance food security.

A viable insurance scheme will undoubtedly act as a vehicle of growth for the small-scale agricultural sector of developing countries of the tropics and subtropics. If for no other reason, idle funds kept aside by the small farmer and other concerned entities as reserves in case of unexpected adverse production circumstances can now be profitably employed.

Chapter Eight:

SMALLHOLDER DISASTER-RISK REDUCTION

Justification

Boundaries set for this book preclude in-depth analyses of the adverse effects of climate change and hydro-meteorological events such as floods, droughts and wildfire. Human-induced crises including environmental degradation, market manipulation and civil strife are also not to be analyzed. Natural *or* such man-made *hazards* (phenomena or events which are potentially damaging to the practice of agriculture) however tend to render some types of agriculture unsustainable. Since these hazards cannot be ignored or wished away, the stakeholders in small-scale agriculture who are consumers, producers or facilitate the production process deserve to have at least a superficial knowledge about ways of managing them.

Concepts

Disaster-Risk Reduction is a major component of Disaster-Risk Management. It essentially entails minimizing damage doneby hazards through a culture of avoidance or prevention since disasters follow in the wake of persistent or severe hazards. Other components of Disaster-Risk Management include; Disaster-Risk Assessment, Disaster Preparedness, Disaster Early-Warning, Disaster Response and Disaster Mitigation (Relief and Rehabilitation).

Smallholder Agriculture Disaster-Risk Management strategies include;

Territorial Planning aimed at climate-conscious land use in order to minimize disaster-risks, protect the environment and maximize agricultural productivity. Instituting sound soil/water management as well as good farming systems and practices are also aspects of Territorial Planning

odifying agricultural production processes and encouraging such practices as crop diversification, crop rotation and agro-forestry for increased protection against hazards.

Creating a conducive atmosphere for the provision of low interest credit and risk-transfer instruments, such as agricultural and personal insurance.

Reassessing physical infrastructure to ensure that they are climate-proof, taking into account new and recurring phenomena such as perennial flooding due to intensified rainfall and rising sea-levels.

Improving Local Governance and sensitizing other social groups in order to conserve the environment and its resources, identify and facilitate access to improved agricultural practices, inputs and services, finance and markets. The enhancement of disaster preparedness and recovery programmes through social networking is also envisaged.

Providing more efficient and effective data-collection, analyses and climate/weather prediction stations as well as constantly improving early-warning systems.

Providing adequate as well as timely disaster-response and rehabilitation mechanisms through the introduction of social safety-nets and coordination of post-disaster relief efforts by concerned parties.

Developing the human infrastructure capacity and capability required to implement all the strategies mentioned above.

Disaster-risk management strategies are generally skewed towards the amelioration of natural hazards (climatic and hydro-meteorological events). This is understandable given the low level of predictability associated with human-induced hazards where the strategy applied is basically to be prepared for their possible occurrence.

Partners

Disaster-risk management involves a wide spectrum of players ranging from the potential or actual victim of the hazard to civil society groups, the affected community, state/provincial governments and their organs, national/federal governments and their subsidiaries, private corporate and faith-based entities as well as international aid and humanitarian organizations.

Disaster-risk reduction strategies in small-scale agriculture revolve around the twin objectives of increasing agricultural productivity and making farmer-incomes less vulnerable to disasters. The various activities of the partners involved in the fine-tuning and implementation of these strategies are best coordinated on community bases. At this level, the peculiar requirements of each locality can be taken into account. Common to every locality, however are specific roles expected of the government and the smallholder.

Government Roles

What is expected of governments and their agencies to facilitate Disaster-risk reduction includes the following;

Preserving and providing climate-smart, improved seed-stock. These would include heat, drought, disease and salt-tolerant
varieties.

Providing integrated water-resource management systems (water-shed protection, protection of wet-lands, reforestation, etc).

Providing community-based processing and post-harvest seed, feed and fodder storage facilities.

Facilitating access to subsidized credit and risk-transfer instruments.

Improving collection, analyses and dissemination of climate /weather prediction, hazard early-warning and current marketing information.

Strengthening extension and advisory services on climate, pest / disease risk reduction.

Establishing and managing cooperation between local groups, government and other agencies in order to avoid duplication of efforts and increase effectiveness.

Built or reconstruct rural physical infrastructure to cope with climate/weather-related hazards.

Some regions of the world have above-average Natural Disaster-risks. For instance, regions within the Inter-Tropical Convergence Zone are prone to landslides, hailstorms, flash-flooding and thunderstorms. Other areas located on geological fault-lines are prone to earth tremours.

Human activity induced Disaster Risks are higher for farm sites located within the vicinity of large industrial complexes such as crude oil Drilling Rigs.

It is the responsibility of government to establish adequate weather-monitoring and disaster early-warning systems. Government also needs to put in place and enforce strict industrial safety regulations.

Governments of developing countries should be keenly aware that the extent to which smallholders react positively to its strategies for Disaster-risk reduction will largely increase with higher levels of what can be reasonably termed *good health*, disposable income, education, access to relevant information (enlightenment) as well as the assurance of safeguards for lean and post-disaster periods.

Smallholder Roles

Disasters result in lowered production for smallholders. Disaster Mitigation involves complex, planned and coordinated processes which require considerable resources over a long period of time. Many small farmers in developing countries have little or no savings or easy access to formal credit. They usually do not have insurance or social safety-nets.

Reflex responses to disasters by small farmers usually consist of; relocation, seeking for grants and loans from informal sources, engaging in craft (carving, weaving etc), depending on surviving root-crops and wild foods (game/plant forage), eating less or less-nutritious foods, selling surviving small livestock (chicken, goats, etc), engaging in auxiliary occupations (hunting, fishing, gathering, etc) or temporarily working as wage-labourers.

The above measures are however inadequate to cope with the scale of most disasters. It is therefore imperative that smallholders implement Disaster-risk reduction measures since it is in their best interest.

Disaster-risk reduction strategies for the smallholder engaged in tropical agricultural production include;

> Adopting climate-smart and disease-resistant seed-stock.

> Adopting improved farming systems that also promote soil conservation such as; mixed (plant/livestock) farming, agro-forestry (inter-spacing agricultural plants

shrubs and vegetables with tree crops), reduced land-tillage, increased use of manure
(disease-free crop residue/animal droppings), contour farming, terracing for sloppy terrain, minimizing *slash and burn* methods of land clearing, use of live barriers (tree-lines), mulching and crop-rotation.

Adopting water-conservation techniques such as efficient irrigation systems (including drip-water irrigation), roof-top and communal rainwater harvesting as well as shade-nets to reduce evapo-transpiration where feasible.

Crop diversification with combinations of crops not affected in the same way by different natural hazards or market shocks.

Avoiding hazard-prone locations such as low-lying flood or coastal plains and constructing dykes where the use of such land is inevitable.

Climate/weather-prediction data should be used to schedule production activities. (Weather forecasting enables a switch of crop variety or changes in the planting calendar when necessary)

Organizing themselves into formal or informal groups to enable or increase access to improved production techniques/technologies, markets, storage facilities, formal credit, agro and extension services as well as insurance. Groups are also able to attract attention and access aid faster should the need arise.

Chapter Nine

SMALLHOLDER PRODUCTION PLANNING

Back to Land

In recent times, many developing countries have had the misfortune of finding themselves heavily indebted to domestic and foreign creditors. As they struggle to make the mandatory payments, they invariably make some sort of structural adjustments to their economies. Countries which have a large component of their debt portfolio held by external creditors try to reduce their level of dependence on imported items. Luxury goods and agricultural products usually top the list of items short-listed for import restriction. Sudden import restrictions in an economy hitherto dependent on agricultural imports lead to shortfalls in the domestic supply of food and fibre for consumption and industrial purposes. In order to make up for this shortfall, a deliberate policy of stimulating agricultural production is embarked upon by policy-makers directing the affairs of governments in these countries.

Periods of structural adjustment are characterized by stringent domestic monetary and fiscal policies prescribed by foreign creditor and donor institutions in cases where developing countries are heavily indebted. These policies tend to create a conducive environment for increased productivity in order to boost the Gross National Product. More individuals, groups and corporations get involved in agricultural production while farmers are encouraged to increase their levels of production.

Since structural adjustment periods are typified by

high rates of inflation, users of agriculture-based inputs for production processes embark on farm projects to reduce their costs and ensure continuous supply. People also take to agriculture in order to lower their food bills and augment their incomes, while others now find agricultural production lucrative.

Planning smallholder agriculture implies a conscious effort to influence the *normal* process of agricultural production. The normal process of agriculture in tropical developing countries is predominantly near-subsistence or peasant-based production with an almost-total dependence on providence and the benevolence of endemic pests, diseases and weather elements. Planning entails a deliberate attempt to depart from this norm by the adoption of modern techniques and the acquisition of more resources which are utilized in an efficient manner in order to increase output and productivity.

Potential small-scale producers have been known to underestimate the complexity of agricultural processes. Agriculture is erroneously believed to be a simple occupation or even hobby, open to everybody regardless of one's relevant experience or lack of it. It is often opined that if the typical *illiterate rural farmer* is able to make a living from agriculture, anyone with a significant level of enlightenment should be able to perform at least, equally well. Efforts at actual production may however lead to greater respect for the farmer and an appreciation of the fact that the illiterate farmer has acquired a wealth of know-how through decades of relevant personal experience and knowledge passed onto him by fore-bearers.

Relevant experience may be the decisive factor between success and failure in smallholder agricultural production. Fortunately, this experience does not necessarily have to be personal but can be acquired through other means. Sufficient impersonal experience acquired from a variety of credible sources with thorough planning, can give an intending or new farmer an edge over an operating or old farmer. This possibility arises because of the dynamic nature of agricultural production cycles. There is the overarching phenomenon of Climate Change. Species of plants and animals as well as pests and disease-organisms evolve and adapt to new circumstances over time. New technologies and techniques of production are also introduced from time to time.

Ability to adapt to changing circumstances and avoid mistakes made by other farmers in the past, is the key to profitable production. The successful farmer is usually the one with access to knowledge about the most efficient and affordable techniques suited to his peculiar production circumstances. He/she also makes use of improved and proven varieties of plants and animals.

Requisites

The decision to go into farming should be made after considering many factors besides finance. Success in agricultural production hinges on precise planning. Any intending farmer wishing to embark on small-scale production should have sufficient and sustained interest in the project.

Ideally, agricultural production should not be embarked upon as a fad or pastime. The potential producer must have sufficient enthusiasm to see the project come to fruition despite obstacles thrown up by the production process.

It is hardly neither practical nor economical in small-scale production to consult with experts in order to discuss every important decision to be made. A cost-conscious potential farmer must therefore have a working knowledge of the project at hand, acquired from competent and reliable sources such as agricultural institutions and research agencies, successful farmers, extension agents, consultants and input-providers. He should also have easy access to his information-sources in order to clarify confusing or complex issues as the need arises.

A considerable amount of physical exertion is involved in small-scale agricultural production processes. The routine search for relevant information as well as the rigours of actual production dictate that the intended producer should ideally be physically fit and enjoy what can be reasonably termed *good health*.

With all other necessary things in place, the size of the project largely depends on the financial resources that the intending farmer is able and willing to commit to it. He has to consider his sources of agricultural finance, both personal and external. Personal sources consist of savings and current income while external sources include grants and loans. He subsequently has to determine the level of funding that is available from all these sources as well as the possibility of obtaining additional finance should the need arise after commencement of the production process.

Other Requirements

Where the intending farmer is certain that he can cope with the physical demands of agricultural production and is able to muster a reasonable amount of financial resources, he has to find a location most suitable for the project in view

Even though the farmer may have a preferred plant/animal species, he must nonetheless ensure that the project is technically feasible. That is, he has to determine the range of projects that can be undertaken at any of his recommended sites, considering their ecological and other conditions. Technical feasibility concerns here include; climate/weather, soil-characteristics, topography, water supply (irrigation or rainfall), vegetation, disease-incidence and pest prevalence.

An assessment of past and existing agricultural projects around any site under consideration will be most helpful. This assessment provides a range of plants and animals from which choices can be made. Consequently, the intending farmer is able to determine whether his preferred plant or animal species will do well in his preferred location.

The intending farmer considers farm-site accessibility, security for farm assets, proximity to such infrastructure as power and communication where they are indispensable for the production process. Off-grid power and other systems such as solar, wind as well as geostatic satellites are opening up remote locations hitherto considered to be inaccessible. The farmer also considers the sources and proximity to such inputs as fertilizers, pesticides, planting materials and animal stock, in addition to hired labour and agro-services in such forms as mechanized implement-hire, extension, veterinary and pest control services.

A crucial factor in the choice of farm-site is the *social acceptability* of the project. The project must be compatible with community structures, norms and beliefs. It would be considered foolhardy to establish a piggery farm within a predominantly Moslem community since they abhor contact with or the consumption of pork and products contaminated by pigs.

An intending farmer should give serious attention to the availability of facilities for the preservation and storage of his products prior to disposal. He should also ensure that there is easy proximity to assured markets.

Not to be left out is the adaptability of the farmer's family to new circumstances if relocation is inevitable. Such issues as the attitude of locals towards new settlers, proximity to appropriate educational and medical facilities, personal security and the availability of social, recreational, religious, law-enforcement and other amenities are to be considered.

System and Item Definition

After working out technical matters, an intending farmer has to decide on the system of production to adopt. The most efficient method of production, subject to crucial factors, has to be worked out. For instance, does one compound animal feed or depend on commercially available varieties? *System definition* encompasses planning for risk and uncertainty where control measures such as diversification, flexibility in equipment, cost and timing, mixed farming or mixed cropping could be incorporated.

Item definition involves further selection within broad choices. For instance, a farmer who opts for cereal production has to make a decision among options such as sorghum, maize or millet. Opting for poultry production leads to first choosing between intensive or extensive production systems, cage or deep-litter (where an intensive system has been chosen) and selecting among the poultry species notably, domestic fowl, turkey, duck or guinea-fowl. Even within the domestic fowl category, a choice has to be made between cockerels, broilers and layers or any combination of the three categories.

An intending farmer who wishes to make the most efficient and effective use of available resources has to compile a list of equipment, manpower requirements, facilities and activities necessary to produce a stated quantity of the produce in view. It is also important to find out the ideal spacing requirement for plants species or ideal capacity for animal species which produces optimum results. This way, overcrowding of the species or under utilization of land and facilities is avoided.

Next, the equipment and facilities, volume of inputs and magnitude of services required to sustain the reference level of output need to be quantified and given approximate values based on careful analyses. Chosen output levels could be, for instance, 1000 broilers or 50 metric tonnes of maize per production cycle. In the case of maize production, the list would include such items as land requirements and preparation, equipment specifications, labour requirements, material inputs such as fertilizers, pesticides, seeds, hoes, machetes, etc. For broiler production, the list would include fixed and semi-fixed capital items such as sheds, (cages are optional), feeders, drinkers etc. The broiler project list would also include day-old chicks, feed and medication. Provisions also have to be made for utilities (water, sanitation), staff requirements, pest control and veterinary services, security, facility maintenance and equipment replacement.

Valuation and Assessment

After System and Item definition, the farmer has a list of requirements to initiate and sustain the production process. Quantities and monetary values now have to be assigned to these requirements. Valuation of goods and services to be utilized in future time periods is approached with caution because prices are hardly stable in the usually turbulent economies of developing countries. An intending farmer has to undertake a cursory survey of past and present prices of needed goods and services around his production location in order to establish a trend for the cost of each item. This will enable an enlightened forecast of future price movements which will affect his overall cost outlay. He also has to forecast a price range for each anticipated product. These forecasts are made with specific dates as reference points. The results of these forecasts only remain valid when actual production is carried out within the stipulated time frame.

In assigning costs to inputs, it is prudent for an

intending farmer to slightly inflate item costs. He may alternatively decide to add a small percentage (say 5%) to the overall cost profile as allowance for contingencies. The advantage of inflating costs is that slight deviations from budgeted costs do not adversely affect the production process. In like manner, the farmer could just under-estimate the anticipated product prices. The primary consideration is to ensure the availability of resources needed to carry the seed-stock to the point of harvest in spite of adverse cost variations. Increased production costs are normally compensated for by higher product prices. The project may not be derailed by cost inflation in the final analysis if there are enough resources to bring it to fruition.

It is assumed that the intending farmer now has an idea of his cost of production for a specified level of output. The next step is to consider his finances in order to determine the volume of production that he can comfortably sustain. This volume would be roughly calculated as fractions or multiples of his reference level of output. A more rigorous method of determining the ideal level of production may be employed if possible. This is because his overall cost structure is made up of both fixed and variable components. *Fixed costs* are costs which do not change in the same magnitude with variations in the output level. In the case of broiler production these include physical installations, utilities, security and salaries. (The same amount of electric power or fuel is used to light up a poultry shed whether it contains 100 or 150 broilers. The poultry attendants' wages do not also vary with the number of broilers per shed). *Variable costs* change in the same direction and roughly, the same magnitude with changes in output levels. Variable costs include feed and medication. Allowances should be made for the peculiarities of each type of cost when the farmer is computing his ideal volume of production. The obvious next step is to compare *total projected cost* with *total projected revenue* per production cycle. The intending farmer should not base revenue calculations on 100% of his anticipated output. He should make some allowance for possible losses. If part of the output is to be devoted to family consumption, it should also form part of the revenue calculations. Where the production cycle involves an appreciable time-span, it may be necessary to factor in the *time-value* of money using acceptable discount rates, so that a more realistic assessment may be made. This is because a unit of currency today may not have the same real value next year as a result of inflation, official devaluation, etc.

In a broiler or maize production project, comparison simply involves matching sales revenue from anticipated output with total cost incurred in a production cycle.

For an egg production project, revenue accrues over a relatively long period of time (more than one calendar year.) The intending farmer in this case has to compute projected egg sales over the entire laying period of a particular batch of layers as well as the value of the birds as table birds (old layers) at the end of lay. This computation takes into account the laying capacities of the birds during the various stages of their life-cycle. The farmer matches the revenue arrived at above with the cost of purchase of pullets (point-of-lay birds) or day-old chicks plus cost of housing/maintaining the layers up to the point of disposal. Comparison of costs and revenue is carried out in order to determine if the project is financially viable (profitable) and if the magnitude of profit is enough to justify the amount of resources required by the production process.

Action Plan

Where the processes of matching total revenues with total costs leave the intending farmer satisfied with the level of anticipated profit, he draws up a *Plan of Action*. His action plan consists of a schedule of activities to be performed at specific instances during the production cycle.

The schedule for an intending farmer anticipating rain-fed maize production could be drawn up as follows:

Date	*Activity*
XXX TO XXX	LAND CLEARING, TILLING / PLOWING
XXX TO XXX	PLANTING
XXX TO XXX	PESTICIDE APPLICATION*
XXX TO XXX	WEEDING**
XXX TO XXX	FERTILIZER APPLICATION***
XXX TO XXX	HARVESTING/DRYING/HUSKING/STORAGE
XXX TO XXX	SALES/DISPOSAL

*Pesticide application may be replaced by cultural practices aimed at the control/eradication of pest infestation.

**Frequency of weeding depends on type of vegetation.

***Fertilizer application may be carried out at the planting stage, if mechanization is employed.

Depending on the species, maize has a definite growth cycle. The exact dates for various activities are hardly ever fixed beforehand in rain-fed agriculture for obvious reasons. The farmer should be able to determine time-periods within which each activity is to be carried out if he has done enough background research. For instance, rainfall follows largely predictable seasonal patterns, depending on the location.

Of recent, there have been severe weather fluctuations on a global scale. The science of weather forecasting has fortunately been steadily improving. Accessing information about present and future weather patterns has also become easier via the use of smart phones with internet access. Farmers into crop production, apart from taking a look at the weather patterns in the recent past, should also acquaint themselves with future weather expectations before committing to actual production.

The schedule for egg production could be as follows.

Date	*Activity*
XXX TO XXX	CONSTRUCTION / EQUIPMENT INSTALLATION.
XXX TO XXX	STOCK ACQUISITION: (CHICKS/POINT-OF-LAY PULLETS)
XXX TO XXX	VACCINATION PROGRAMME
XXX TO XXX	MAINTENANCE*

| XXX TO XXX | EGG SALES |
| XXX TO XXX | SALE OF OLD LAYERS |

*Maintenance entails the housing, feeding and medication of the birds from the date of acquisition to the point of disposal.

The schedules of activities shown above are simplified versions. In a functional schedule for egg production, the various vaccinations, other necessary activities such as de-worming and optional activities such as de-beaking are listed as separate items. The feed types utilized at the various stages of growth, are also listed as separate items.

Activities required for animal production projects are more amenable to precise planning than for crop production. This is because intensive animal production is carried out within a fairly-controllable environment. For instance, the domestic fowl can be reared at any time of the year, regardless of prevailing weather conditions. Nevertheless, table-birds such as broilers and cockerels should be timed to fully mature at periods of peak demand (festival periods) especially where refrigeration or cold-storage poses a formidable problem.

Using his Plan of Action as a guide, an intending farmer is able to determine approximate moments in time when he requires specific amounts of cash. He subsequently evaluates his sources of finance in order to make sure that cash will be available when needed and in sufficient amounts, especially if he is dependent to a large extent on external funding.

Tests

Having gone through the stages enumerated above, there are *quantitative analytical tests* that can be performed to determine the financial viability of a project. These tests include *Net Present Worth, Internal Rate of Return, Benefit/Cost Ratio* and *Weighted Average Cost of Capital.* The relevance of these tests to the smallholder is that institutional providers of credit use these and similar tests to evaluate projects seeking for funding. A more detailed discussion of these tests is beyond the scope set for this book. Readers may choose to familiarize themselves with these tests if they consider its knowledge indispensable.

Some of the tests mentioned above can be carried out by numerically-inclined, intending farmers who are familiar with discounted cash flow analyses. Alternatively, intending farmers who are not disposed towards quantitative analyses could seek the assistance of people with the requisite knowledge to perform the suggested tests for them. Such people include bankers, teachers and students of accounting, economics, banking and finance. The intending farmer may limit himself to simple *Sensitivity Tests* which he can perform by himself. He may choose to find out for instance, how sensitive the project is to a fall in product price by a certain percentage-point. He does this by calculating whether he will still *break-even* at a selected lower price level. (*Break-Even Point* occurs where his *Expected Total Revenue* just equals his *Expected Total Costs.*) To do this, he chooses a percentage of his expected unit product price, for instance

80% of his expected selling price per unit of product and uses the new price to calculate his expected total revenue. He matches this new total revenue with his original expected total cost to see whether he will still break-even or even record a significant profit margin. Depending on his level of optimism and the lower the percentage of his original unit price at which he breaks even, there is stronger recommendation for the project, since it is able to withstand a significant increase in production costs or decrease in product prices.

There are other criteria besides financial viability that could justify the execution of a project depending on the whims of the entrepreneur. Motives such as extracurricular activity, personal interest including political ambition, retirement option, social responsibility and philanthropy may be paramount to the investor. People may be loath to abandon a project which fails to meet the profitability criterion after spending much time and effort on its planning. Such people may even be optimistic about its future viability. An intending small-scale farmer should follow the maxim that *it is better to be safe than sorry*, especially if he cannot easily absorb losses.

Embarking on research or consulting with experts will unearth potentially-profitable, agricultural projects available for exploitation. Where institutionalized risk control measures like agricultural insurance exists, it would be a wise precaution for the smallholder to take advantage of them.

The processes involved in production-planning may appear overwhelming to the typical smallholder with limited educational qualifications and low level of enlightenment.

Group Action aimed at harnessing the talents of available local technocrats to assist the small farmers in their individual capacities, is an option available for exploitation.

Intending farmers should not rush to purchase off-the-shelf feasibility studies from consultants or agriculture-specialists. At best, these well-crafted plans may enable a farmer obtain formal production credit. It will however be of little use to him/her in the course of actual production. The farmer and the expert should jointly determine projects that are viable based on the realities on ground. Where the farmer's preferred project does not pass the viability test, it should be dropped for another option.

Chapter Ten:

AGRICULTURAL PROJECT FAILURE

Failure

As is usual with production ventures, agricultural projects have their attendant problems. The uniqueness of agriculture lies in its significantly higher levels of risks and uncertainties when compared to mainstream industrial projects. This phenomenon could be mainly ascribed to the fact that agriculture involves living things which are largely conditioned by nature and the environment. The chances of failure for agricultural projects are consequently higher than for projects involving non-living things.

For our purposes, an agricultural project will be deemed to have failed if it does not satisfy the motive for which it was set up. Motivation for initiating agricultural projects could arise from short or long-term goals, among which are extracurricular activity, personal interest, social responsibility and philanthropy. Another possible motivational factor is profit, which we will here adopt as the primary goal for the initiation of small-scale agricultural projects. *Profit motivation* implies that the farmer wishes to obtain as much incremental returns as possible from his investment in a particular agricultural project.

Successful agricultural projects have been known to fail as a result of a sudden inability of owner-operators to cope with the physical exertion demanded by the production processes. This lack of ability to cope may result from physiological disability or ill-health. A farmer's dependant may become afflicted with a lingering or recurring ailment which also serves as a formidable source of distraction for a healthy operator. Sicknesses arising from epidemics may also limit the productive capacity of the farmer's labour force. In the extreme case of a farm-operator's death, his survivors may have neither the ability nor enthusiasm to stay in production.

Delegating policy initiatives and supervision of production and/or marketing activities to less-knowledgeable or less-enthusiastic individuals could herald the failure of an agricultural project.

There could be willful neglect or apathy among some farm employees who are not convinced that their relative economic wellbeing would change substantially, if the fortunes of the project were to improve.

The relocation of an owner-operator to a place far away from his farm-site could lead to unintended neglect and subsequent failure of a small-scale agricultural project. The supervision of the project may constitute too much of a strain, especially when the farmer is also involved in other time and energy-demanding pursuits.

Weather elements and physical environments which adversely affect production could be causes of project failure. Low yields could result from soil erosion, flooding, droughts, endemic diseases and pests among other possible natural hazards.

Allied with the above are failures caused by human activity such as civil strife, fire outbreaks and industrial accidents resulting in environmental pollution. Crude-oil spillage and soil-contamination by entities involved in mining activities are typical examples. There could also be gross abuse of land and water resources heralding the failure of agricultural projects within the immediate environment.

Failure to plan is a major operator-induced cause of project failure. As the saying goes, *"who fails to plan, plans to fail"*. Where planning is undertaken, the quality of the project plan determines to what extent, success is achieved. Features of bad agricultural production plans include; wrong choice or bad timing of projects, making inadequate provisions for the continuous supply of needed inputs and agro-services, poor and ineffective storage and marketing arrangements as well as inadequate strategies for managing risk and uncertainty.

Negligence on the part of the owner-operator could result in project failure. Negligence occurs in such forms as unnecessary deviations from schedules, failure to keep track of overall farm performance through comprehensive and up-to-date record-keeping. The result of instances of negligence could be pest and disease incidence, destruction of farm assets, under-utilization of labour, wastage of material inputs as well as substantial losses due to theft.

Cash-flow problems occur as a direct consequence of bad planning and are a significant cause of the failure of smallholder agricultural projects. Cash shortages occur as a result of the farm operator's inability to anticipate his entire cash needs and make adequate provisions to ensure timely and sufficient cash inflows. Shortages also occur when the operator is unable to raise enough cash to cover unforeseen expenses. Operators who rely on credit financing are most susceptible to cash shortfalls. Diversion of funds meant for production purposes usually results from failure to separate business accounts from personal income and expenditure accounts. This may lead to cash shortfalls. Unplanned and overambitious expansion of project size can also lead to cash shortfalls.

Adverse micro or macroeconomic circumstances could lead to project failure. For instance, cost reduction measures by neighbouring farmers, such as the introduction of more-productive varieties of plants and animals or more efficient production techniques, may lead to reduction in local market prices. This could render a farmer's production venture unprofitable. The emergence of a nearby large-scale producer, who enjoys economies of scale, could also drive smaller farmers out of business. Reduced demand for specific products as a result of global changes in income, tastes or fashion, may force farmers to abandon their projects. A market glut for specific products could lead to market prices being lower than production costs for an extended period of time.

Politico-legal causes of project failure include cessation of land tenure or loss of substantial assets as a result of legal arbitration or government acquisition. Government policies such as liberalization of imports of food, fibre and processed agricultural products may provide cheaper alternatives to local products and reduce demand for local products to the extent that it is no longer profitable to remain in production.

Smallholder projects may fail under the circumstances enumerated above when their contingency plans do not cover such scenarios.

Survival Strategies

No agricultural project should be considered too small or simple to implement strategies aimed at facilitating its continued existence and profitability. Survival strategies, prudent management and an element of good fortune are what it takes to ensure the evolution of a farmer from small to larger scale.

Farm operators should always be conscious of their production processes. Even where an in-depth knowledge of a particular process has been acquired, it is worthwhile to remember that agriculture is always in a state of evolution. New strains of plants, animals, pests, diseases and production systems appear over time. Operators should therefore update their knowledge frequently. The smallholder should not hesitate to solicit for help from more experienced producers, cooperatives (avoid direct competitors if possible), consultants, extension personnel, input suppliers, academic and research institutions as the need arises.

Projects chosen by potential smallholders should be generally known to be profitable and the project plan should be subjected to some form of analysis. Tests to determine the financial viability of smallholder agricultural projects should be performed where feasible. These tests should include sensitivity tests to determine the impact of adverse changes in prices or costs on the project outcome.

In seeking specialist advice and professional services, a potential smallholder should deal with capable people.

Some small-scale project plans or feasibility studies purchased from emergency agricultural consultants never got off the drawing board. Instances abound where projects failed as a result of retaining the services of incompetent or quack animal-health consultants, use of unproven techniques and technologies, removal of topsoil and destruction of soil aggregates by inexperienced land-clearing agencies or destruction of seedlings as a result of improper or uncontrolled application of agrochemicals based on the recommendations of incompetent salespersons.

Agricultural projects selected for smallholder implementation must be suitable or adapted to the physical environment in which they are to take place. Factors to be considered in the location of such projects should include the nature of the terrain and soil characteristics, endemic pests and diseases, susceptibility to erosion and flooding, general security for plant and products, input and labour availability, access, markets, essential utilities and expansion possibilities.

River basins were hitherto considered premium locations for arable farming. Of recent, drastic changes in weather patterns have led to restricted or excessive moisture resulting from insufficient or excessive rainfall. Unscheduled releases of water from overflowing dams have become more frequent. Smallholders operating wthin river plains should as much as possible select hardy and/or flood-tolerant plant species with shorter maturation cycles. This shortens the production period thereby reducing the magnitude of risk and uncertainty ascribed to possible flooding. The provision of water-drainage channels should also be an integral part of the physical layout of every farm holding.

Allied with above, is the need to make use of weather forecasts and to access long-term meteorological reports in order to increase the chances of successful production of crops with low-tolerance for weather deviations.

Intending farmers should avoid the acquisition of lands situated along the migratory path of nomadic herdsmen, lands in dispute (between family-members, villagers, communities) or lands whose owners have unsavoury reputations. This will serve to minimize costly distractions and possible loss of assets. On the same token, it would be wise for the farmer to sort out his pressing socio-cultural and other pertinent issues serving as formidable sources of distraction.

Depending on the religious inclination of the farmer, he may wish to identify and tackle spiritual issues which may hinder the successful execution of the project all other necessary conditions having been met.

Where feasible, small farmers should negotiate for reasonably long leases from landowners if permanent structures are to be erected, late-maturing species are to be cultivated or considerable improvements are to be made to the land. This reduces the chances of ejection before the full potential of the project is realized, especially where the value of leased land tends to appreciate rapidly over time.

Risk Management as a survival strategy aims at nullifying or minimizing the adverse effects of unfavourable production circumstances. Specific measures that can be employed by the smallholder include; diversification of products, flexibility in equipment and timing (purchases are made at the last possible moment to enable a quick switch from one project type to another if it becomes necessary), choice of more reliable projects at the expense of greater profit, contract-pricing and agricultural insurance. Note that options for planning for risks and uncertainties are not limited to these measures.

Cost-control stands out as a survival strategy for the small-scale agricultural producer. This entails expending as little resources as possible in the production of a unit of output without compromising its quality. Cost-control measures are initiated even before the start of actual production. Equipment not often put to use could be hired instead of purchased, critical inputs could be acquired at the least cost in the off-season when little agricultural activities take place, (if the operator can afford to tie down funds). Collaborative efforts could be explored for the procurement of inputs while cheaper production alternatives should be chosen; for instance compounding feed and fodder as against buying commercially available brands.

Smallholders should continually strive to promote efficiency as a means of minimizing unit costs. High-yield varieties coupled with least-cost production methods are sure ways of increasing efficiency. Such incentives as reward for hard work and sanctions for wastage and theft among farm workers also go a long way in entrenching efficiency in a smallholder agricultural project, thereby enhancing its chances of survival.

Cash sufficiency increases the chances of survival of a smallholder agricultural project. *Cash sufficiency* implies having enough money to initiate and complete the production process. This also implies having arrangements in place for extra cash in case of emergencies, scheduling cash availability so that obligations can be met as they fall due, implementing phased expansion activities and ensuring that there is enough cash to cater for the family's domestic and other needs during and immediately after a production cycle.

Some basic inputs required for smallholder agricultural projects such as agrochemicals could be in short supply during the production-cycle. The success of improved production systems and varieties are usually tied to adequate supplies of agrochemicals. A wise farm operator would ensure that he has guaranteed access to an adequate supply of fertilizers, herbicides, pesticides, vaccines and other animal health-care products as the case may be, before embarking on actual production. In addition to having an adequate stock of agrochemicals, farmers should avoid the use of prohibited, expired or obsolete products for purposes of agricultural production, preservation and storage.

Small-scale producers should make some form of arrangement no matter how rudimentary, to at least partially process their products in order to increase its shelf-life,. By so doing, they may not be forced to dispose of their products at any ruling market price for fear of deterioration. Partial processing also increases the value as well as improves the storage and haulage of agricultural products.

Visits to commodity markets used by small-scale producers confirm that a significant amount of produce goes to waste at the end of each market day. Sellers within these markets will not lower prices or increase volumes in order to increase demand, usually in deference to market-union directives aimed at creating artificial scarcity. Small producers who sell directly to consumers should exercise discretion in times of unfavourable market situations. Like the popular adage says *"a bird in hand is worth two in the bush"*.

Generally, smallholders should widen their customer-base as much as possible. They should avoid situations where a few major customers buy all their produce. This will ensure that they do not encounter problems of how to profitably dispose of produce if any particular customer fails to show up.

Lastly, the small-scale agricultural producer should strive to build and maintain a good reputation for his products. He could achieve this by constantly improving the quality of his products in terms of packaging, cleanliness, freshness and freedom from contamination, pest and disease infestation. He should also maintain a good public image by refraining from dishonest practices. When a smallholder is known to be reputable, he will hardly ever lack customers for his products.

Chapter Eleven

GROUP ACTION

Farmer-Groups

In all spheres of human endeavour, individuals on the average, tend to seek out others with similar interests to facilitate the attainment of their personal goals through the satisfaction of needs common to all. Individuals with similar pecuniary motives may likewise come together in some form of association. They do this in the belief that their individual and collective interests could be taken care of through group effort. Associations based on economic considerations spring up or attain prominence with each passing day. This is an eloquent testimony to the fact that group undertakings are better at solving member's problems than their individual efforts.

Groups or associations of smallholder agricultural producers exist in many developing countries. The proportion of small farmers aggregated into groups are however small in relation to the total number of smallholders. This phenomenon is partly attributable to the fact that ideal conditions for the formation of farmer-groups do not exist in these countries.

The fundamental condition for the initiation of group action by small farmers is the awareness of unjust policies aimed at them or some options that are definitely out of their reach as individuals. This awareness fuels the desire to ameliorate the existing situation by exploiting the maxim that *"two heads are better than one"*. Group action among small farmers denotes a situation where they decide to come together to work for mutual benefit.

Cooperation among smallholder agricultural producers usually occur in two forms; individual farmers who are involved in some form of group enterprise and individual farmers with similar but separate enterprises coming together in a loose type of horizontal integration. In the second form, the farmers take their independent production decisions but cooperate in finding solutions to such problems as the acquisition of inputs and/or the disposal of products. This form is more likely to occur in free-market economies where production efforts are not primarily under strict government supervision. Our focus will be on the second form of smallholder cooperation.

Smallholder cooperation may be formal or informal. The formal associations are known as *Cooperatives*. Farmer cooperatives are duly registered, enjoy government recognition and are guided by rules and regulations which must be strictly adhered to. Informal associations on the other hand are less rigid. They are unregistered and do not enjoy government or legal recognition as trade groups. They also operate under rules jointly formulated by members in order to suit their purposes, which may not follow principles adopted by formal groups. Some informal groups known as *pre-cooperatives* are however formed with the intention of evolving into formal groups with the passage of time.

Group action among smallholders could be for the purpose of solving specific problems such as credit procurement. It could also be for the purpose of simultaneously solving a variety of problems in such areas as input or service procurement, marketing, storage and transportation. The majority of smallholder cooperation is however product-specific. That is, they usually revolve around specific items within such product groups as cereal-grains, farm animals and cash crops.

Problems

Group action as earlier mentioned arises from an awareness of felt needs which cannot be easily fulfilled as individuals. The insignificant number of farmer-groups in developing countries is directly related to the fact that they possess predominantly *peasant-farmer* agricultural sectors. Since farming at near-subsistence level does not overwhelm the smallholder, he has little need for group action. As a peasant / subsistence farmer is increasingly exposed to new and external influences and consequently desires to increase his output in order to enhance his living standards, he encounters new problems which are entirely outside the scope of his abilities. One is hardly able to list all the production-related problems facing a smallholder but a few problems where group action can be applied will suffice.

The first major problem encountered by a peasant farmer evolving from subsistence to small scale or transforming from small-scale to something bigger is finance. Additional resources are required with which to modernize and increase production. Since he is not influential or even well-off even in relative terms, he is unable to obtain easy loans from both formal and informal sources.

The magnitude of inputs required by the small-scale

agricultural producer is consistent with the size of his farm holdings. By implication, he cannot qualify for price discounts from input suppliers and is forced to make his purchases at retail prices, instead of cheaper wholesale rates. He is not even guaranteed a continuous supply in times of scarcity since he is not a notable or high-value customer.

The typical smallholder's farm-holdings are usually fragmented and sited in different locations. As such, he cannot save time, money or easily put more land under cultivation by employing mechanization. In addition, mechanization services are not cost-effective for him because of the small cumulative size of his farm holdings. For instance, a farmer with just two or three hectares of land cannot conveniently engage heavy industrial machinery to work on his sites since the cost of transporting such machinery to and fro the sites may far exceed the cost of actual work to be done, which may be paid for on a daily and not hourly basis.

Agricultural production is usually carried out in areas remote from urban centres where most of the consumption takes place, whereas agricultural products are usually delicate, of a bulky nature and may require special handling techniques. Evacuation of products from the farms to the markets is more expensive for farmers who have to make independent transportation arrangements.

Non-processed farm products deteriorate within a relatively short time span after harvest. The smallholder in the absence of processing facilities has to dispose of his products immediately after harvest and he off-loads his products at local market places no matter the prevailing market price.

Smallholders who are kept busy by the demands of the production-cycle and do not usually have sufficient time to update information regularly are often unaware or unable to access markets where they can obtain the most attractive prices.

The individual smallholders cannot dictate or influence product-prices. As result of the insignificant quantity of his output he is often left at the mercy of middlemen who are ever willing to take undue advantage of him.

Most facilitating agencies in smallholder agriculture and other government agencies in the developing countries do not find it cost-effective to render individual services to smallholders. The smallholder is therefore unable to have easy access to subsidized or free services which are usually technical, educational, extension or input-supply based.

Individual small-scale agricultural producers working in rural communities in the developing countries are unable to lobby or impress upon policy-makers and large corporate entities, their dire need for public utilities and social services. This is because their individual contributions to total output are not significant enough to warrant recognition and entitle them to performance-enhancing incentives.

Group Objectives

The problems of individual smallholder agricultural producers cannot be exactly the same in the strictest sense. Their problems may be similar but they usually manifest themselves in different ways and in varying degrees. To streamline the activities of farmer-groups and at the same time accommodate their individual problems, group aims and objectives must exist which will serve as focal points for group activities.

The broad objectives of groups of smallholders engaged in agricultural production would include to:

> arrange for reliable and regular supplies of production inputs, including credit.
>
> reduce cost of production through increased efficiency, cheaper inputs and sources of finance, improved seeds and stock.
>
> continuously educate members on innovations in production technology, processing and storage.
>
> discover and exploit new markets as well as facilitate the maximum utilization of existing ones.
>
> attract and facilitate government and other intervention in the provision of facilities and amenities aimed at boosting productivity.
>
> leverage on existing technology in order to enhance communications, knowledge and skill acquisition.
>
> monitor the climate and environment as well as facilitate disaster - management.

Advantages

The main advantage of group-action by smallholder

agricultural producers is *increased bargaining power*. While the individual farmers may not amount to much, they collectively constitute a force which may not be lightly dismissed.

Specific advantages of group-action include the opportunity to plan and execute farm operations better. Through constant interaction, smallholders find it easier to identify and solve everyday production problems as well as find solutions to more complex ones.

There is greater efficiency in the use of labour and farm implements. Farmers do not have to tie down capital in the procurement of all the equipment needed because it could be borrowed from other group members or jointly acquired. Neighbouring farm holdings are also able to jointly utilize mechanization at reasonable cost to each farmer.

The chances of project failure for a farmer engaged in group action with other farmers are considerably less. Other members of a group can identify with a farmer's problem and empathize with him. For instance if a group member becomes physically incapacitated during the course of a production cycle, it is likely that other members will provide assistance in the execution of some farm activities such as weeding, planting and harvesting.

Formal lending agencies are more willing to make production credit available on easier terms to small-farmer groups rather than to individual smallholders. Because repayment is a group responsibility, there is added assurance that loans would be repaid. Groups are also in a better position to satisfy the stringent collateral requirements demanded by institutional lenders. While on the same subject matter, formal cooperatives usually have an edge over informal farmer-groups in the procurement of formal credit.

Unit costs of goods and services required for agricultural production are considerably reduced when small-farmer groups embark on bulk purchasing. These groups possess economic powers on a micro level and can therefore ensure a steady supply of critical inputs in addition to being in a position to negotiate better rates at which they buy and sell, since they are essentially local monopolies.

Government and other agricultural development agencies which supply inputs at subsidized rates find it easier to locate and deal with genuine small-scale agricultural producers when they are in clusters. Inputs usually made available by these agencies include animal health-care products and agrochemicals such as fertilizers, pesticides and herbicides. Depending on how stable the groups are, they may even be considered for credit purchases.

Small-farmer groups are in a better position to demand and/or lobby local authorities into giving them public services and utilities provided, sponsored or financed by government agencies. They are also able to attract the attention of local and foreign development organizations and other production-facilitating concerns.

Extension workers and other change-agencies find informing and educating the small-scale agricultural producer easier when they are in groups rather than when they have to locate and deal with them on individual bases. The fact that their different efforts can now be better coordinated and focused with greater opportunities for practical demonstrations; increase the success-rate of these change agents.

Group Action motivates the smallholders to be more involved in *Agricultural Value Chain* activities. The farmers are now able to take up functions such as primary processing and packaging for the final consumer. These value-enhancing functions enable the group to retain the margin that would have gone to middlemen operating within the sector who usually undertake these indispensable functions.

Group Action in executing activities associated with the Agricultural Value Chain invariably leads to further action in other associated matters as the group engages in backward and forward integration activities aimed at enhancing their fortunes. This progressive cycle results in accelerated agricultural development which lays the foundation for the self-sustaining growth, eventual transformation and industrialization of the agricultural sectors of developing nations.

Constraints

The fact that smallholder agricultural producers with similar problems and strong commitment come together to find common solutions do not automatically guarantee the success of the group venture. Certain constraints highlighted below may hinder the attainment of group objectives.

Majority of small-farmer groups in the developing countries are made up of individual farmers with little or no formal education. As a consequence, it is difficult for these groups to embrace modern and improved farming methods which incorporate record-keeping and financial accounting. Continuous interaction with formal agencies and keeping abreast of developments through the use of modern mass communication gadgets and facilities such as the internet may also be required

It may even be quite difficult for potential members of the group to meet up with basic requirements for membership and embrace the norms and attitudes necessary for the attainment of group objectives.

Small-farmer groups usually cannot afford to employ staff to handle the affairs of the group. Management teams are often selected from group members based on their persuasive skills or group perception of their abilities. Some selected members may not possess the zeal, dedication, stamina or competence to pilot the group to success. Some charismatic members of small-farmer groups have also been known to manipulate the other members and use them for their personal political, social or economic agenda at the cost of achieving group cohesion and objectives.

Pre-cooperatives and cooperatives may be forced to adopt certain principles and practices perceived not to be in consonance with profit-maximizing objectives. Some of the Rochdalian Principles on which formal cooperative activity is anchored are not in tune with the circumstances of the smallholder located in a tropical or subtropical developing country.

Conclusion

There are enough compelling reasons for the adoption of group action by smallholder agricultural producers of the developing countries. In addition, policy-makers in developing countries should encourage the formation of farmer cooperatives. These groups are to be utilized as vehicles for the identification, enlightenment and provision of critical and subsidized inputs and services to a significant proportion of the small-scale farmers.

Group action in smallholder agriculture results in better interpersonal relationships which lead to more cohesive communities with better conflict resolution mechanisms in place. It also lays a strong foundation for similar action in rural community development, which in turn translates to the improvement of social amenities and physical infrastructure. These facilities enhance the quality of life enjoyed by the rural dwellers of the developing countries.

Chapter Twelve

EXTENSION IN SMALLHOLDER AGRICULTURE

Philosophy and Role

The Varsity English Dictionary defines extension as *"the art of extending"*. It further defines the word extend as *"to spread, prolong, stretch-out or bestow-on"*. The phrase "bestow-on" best suits the context in which extension is used in relation to agricultural production. It is not essential to delve into the issue of how the term *agricultural extension* came into wide usage; more important is the knowledge of the philosophy behind it. The philosophy behind agricultural extension is the facilitation of the production of food and fibre through a more conducive production environment. The ideal production environment desired is created by the application of scientific methods to identify significant problems and provide practical solutions for them.

The agricultural extension approach involves assisting farmers find solutions to production, domestic and social problems in their various localities. Extension personnel are constantly with the farmers during the production cycle. They facilitate the exploitation of essential support services and combine it with education. This education employs special teaching techniques which include home and farm visits, group targeted meetings, printed literature, social media, audio and visual programmes, exhibitions as well as method and result demonstrations.

Essential support-services required as complements to agricultural extension programmes are not provided by the extension personnel. The provision of infrastructure and amenities which include rural roads, communication channels, education and health-care facilities, irrigation and storage facilities are provided the government. These amenities are solely provided or in conjunction with other local or international development organizations.

Other support-services complementing agricultural extension include the provision of equipment-hire services, efficient cultivation systems, improved seed-stock, research, animal health-care products and agrochemicals.

In many developing countries, a significant proportion of the populace is involved in the bulk of domestic agricultural production which is basically carried out by rural-based smallholders. Our focus will therefore be on agricultural extension as it affects this particular group of producers.

The main role of agricultural extension is to modify the lifestyle of the rural small-scale farmer. The modification aims at achieving a situation where the typical rural smallholder attains higher productivity levels and enjoys a higher standard of living. The systematic approach employed by the extension agent focuses on stimulating the farmers into adjusting to the ever-changing production environment by acquiring new skills, technologies, knowledge and attitudes.

Specific activities carried out by extension personnel include educating smallholders on how to identify their most pressing problems. They subsequently encourage the farmers to embark on group action in order to find solutions to common problems identified.

Extension agents collect and collate research results

and innovations in production techniques from various relevant organizations. They simplify complex information and procedures and convert them into forms that can be easily understood and adopted by rural, not-too-literate, small-scale farmers. They also carry out practical demonstrations to acquaint farmers with exact procedures, then monitor progress made by those who have accepted innovations on trial bases and encourage them to succeed.

Extension agents are trained to carry out on-the-spot assessments of the rural production environment. They are thus able to provide informed feedback on the mileage obtained on policy measures already in place. They also contribute to the information pool available to the appropriate authorities responsible for agricultural policy formulation.

Of great importance to the rural economy, is the fact that extension personnel are in vantage positions to direct research efforts towards addressing specific constraints which serve as major obstacles to smallholder agricultural production.

Extension agents assist the smallholders to exploit agro-services provided by government and other relevant agencies. The special teaching techniques employed by extension personnel include question and answer sessions as well as practical demonstrations. These techniques lay a firm foundation for an awareness of new technologies, input sources and costs, farm implements, market trends, prices and consumer preferences.

A role that is not expected of extension agents is the coercion or intimidation of rural agricultural producers into accepting innovations. Their operational strategy should be to make farmers aware of more profitable alternatives. The farmers are then left to accept any or reject all. The ability of extension personnel to portray the alternatives objectively could be the deciding factor that determines the farmer's subsequent course of action.

Guidelines

An extension programme should ideally be conceived according
to certain principles in order to be most effective. First of all, the functions of an extension agency are best carried out when it is state-owned, since there appears to be little discernible financial motivation that would induce major private sector investment in agricultural extension. Some sizeable corporate organizations engage in some extension services either targeted at entities which directly contribute to their wellbeing or as social responsibility obligations. These efforts can only service an insignificant proportion of the overall extension requirements of the rural smallholders of tropical developing countries.

Extension agencies must have definite objectives and work towards the achievement of specific targets. These objectives should be consistent with government's long-term policy objectives and should therefore be adapted from the agricultural sector policy covering that period in time.

An extension organization should be capable of coordinating its activities with those of other agencies who directly or indirectly provide the support-services required by the extension programme. Such agencies include local authorities, state governments, philanthropic groups and other development organizations associated with the provision of utilities, social amenities, educational, health-care and other physical infrastructure.

An extension organization should be structured in a manner that allows for a convenient and timely two-way communication process between staff in the field, regional hubs and the organization's centre of operations. The structure should also allow for rapid decision-making. Localization of decision-making by middle-level management should be allowed and encouraged within defined limits.

An extension organization should concern itself primarily with problems emanating from agricultural production processes. These problems usually occur in such areas as farming systems, marketing and storage, input procurement and the procurement of specialized services such as animal health-care and mechanized implement hire. They may also engage in Disaster Management in an advisory capacity.

The extension programme may however widen the scope of its activities to tackle such issues as land reform and gender discrimination where practicable, since these issues may have a direct impact on the attainment of its goals. Extension personnel may also encourage and facilitate the efforts of farmers seeking interventions in tackling pressing issues.

Extension philosophy presupposes that rural farmers perform better and at optimal production levels in situations of an improved knowledge base and physical infrastructure as well as favourable domestic, health, economic and socio-cultural conditions. The philosophy however recognizes the need for prudence in tackling the multifaceted problems of the small farmers. Whereas the extension organization should encourage the identification of non-agricultural problems which impact on the wellbeing of the farmers, it should allow other agencies in whose area of competence these problems occur to provide solutions to the identified problems. For instance, the extension programme relies on public health departments to provide adequate healthcare for farm-families. This strategy avoids duplication of efforts and encourages the conservation of scare resources.

The extension programme should be designed and implemented in such a way as to minimally intrude on the farmer's working hours. The programme should be designed along the lines of the farmer's current interests and should be targeted at solving their present and significant production problems.

In some developing countries like Nigeria, special departments are usually created within the agriculture ministries to specifically supervise agricultural extension activities.

Field Agents

A very important cadre of staff within the extension organization is the *field-worker* or *agent*. This category of workers constitutes agents of change that are in direct contact with the rural farmers. The primary role of the field agent is to bring about changes in the farmer's knowledge, skills and attitudes. These changes are directed at achieving certain desirable goals, chief of which is to impart on the farmers, the desire and ability to improve their personal circumstances through their own efforts.

Field agents are trained to observe and assess farming communities. This assessment is not confined to an agricultural perspective but also employs social, economic, climatic/weather ecological and cultural parameters. Having acquired adequate knowledge and a sufficient database, field agents are able to assist the farmers identify their most pressing problems for which solutions can be found. Finally, they encourage the farmers to make choices from an array of feasible scientific solutions and to attain the correct attitudes towards the acquisition of new knowledge and skills required by any solution adopted.

In order to command the respect of community members, field agents must exhibit an acceptable level of competence. They should also possess a personal disposition amiable enough to earn the liking, trust and confidence of the farmers they interact with. They must not have subscribed to the misconception that rural people are inherently hostile to new ideas. The esteem with which host community-members regard extension agents largely determines the success of extension programmes.

Extension agents are hopelessly outnumbered by the rural farmers in most developing countries. They are also overwhelmed by the vastness of the land area to be covered by extension programmes. Under such circumstances, the extension agents have to make use of local farmers to make the most impact. To choose the right calibre of persons to delegate some important tasks of agricultural extension to, the field agents should be able to identify, select and solicit the help of opinion-leaders within the community. Opinion-leaders chosen by field agents must have utmost confidence in the extension programme and a genuine desire to serve members of their respective communities. They should have attained acceptable literacy-levels and be willing to acquire new skills and impart it to others. They also should; possess such traits as dependability and versatility, have *people* and *conceptual* skills as well as some project-planning experience.

Field agents must exhibit high levels of patience and understanding. The production circumstances from which they are trying to wean the farmers are those in which the farmers are most comfortable. They have developed a high level of confidence in their production systems which have over the years proven to be most effective, given their past production circumstances. If the economic viability and profitability of new ideas are proven beyond reasonable doubt, rural farmers will be quick to adopt the new ideas.

Small farmers who are caught in a vicious cycle of poverty are not in any position, nor have the inclination to gamble with the uncertainties associated with untried ideas. If however the means of implementing new ideas are made available at minimal cost to the farmers, they will have nothing to lose and everything to gain by embracing innovation. Field agents should therefore not be content with lining up novelty, but should give equal priority to the acquisition of finance, material inputs, fail-safe mechanisms as well as creating the right environment to bring the innovation to fruition.

Constraints

As earlier mentioned, the long-term objective of any agricultural extension activity is usually in conformity with, and derives from the country's agricultural policies. In many developing countries, the fortunes of agriculture seem to be influenced principally by factors other than consistent, relevant and functional policy instruments. The actual factors controlling agricultural production in these countries appear to be predominantly political and economic stability, weather, market forces and the corporate policies of international finance and credit institutions, where they are heavily indebted. Leadership changes are frequent with each regime discontinuing the policies of its immediate predecessor. In such situations, long-term objectives even when articulated and understood cannot be easily achieved by the extension organization.

Extension programmes are largely government-sponsored. This government involvement dictates that the extension organization is run under a bureaucratic setup. Bureaucracy in most developing countries has its attendant problems which are transferred to the agricultural-extension programme. These problems include; slow and inflexible communication procedures,, slow decision-making processes, length of service substituting for performance as the primary basis for career advancement, arbitrary transfer of personnel, apathy, cronyism, nepotism and mismanagement.

While agricultural extension programmes require

significant financial and manpower resources, it is almost exclusively financed by government. It is therefore not much of a surprise that a common characteristic of most extension programmes is under-funding. This results in such problems as poor staff remunerations and welfare provisions, staff insufficiency in terms of numbers and training, inadequate supervision, lodgings and transportation facilities, inability to carry out needed practical demonstrations, to mention a few. These problems lead to low staff morale and performance.

Educational and research agencies are invaluable to the extension programme. These centres of innovation may be poorly funded and usually operate in isolation from each other. They may also be located in places remote from where the extension activities take place. In such situations, coordination of activities and information exchange between the agencies and extension network is difficult. Research efforts may also be duplicated leading to the uneconomic use of available resources and results may not be transmitted to the farmers as soon as they become available.

Government and Self-Help

National, state and local governments in developing countries have prominent roles to play in the success of agricultural extension programmes, since extension is almost an exclusive government affair. In order to enhance the performance of the scheme, the central government must first of all overhaul its policy formulation and implementation strategies. A definite path or schedule of activities based on specific long-term objectives should be drawn up for agricultural extension. Legislation should be sought in a bid to ensure continuity of programmes and ensure that deviations from schedules do not occur at the whims of ego-centric individuals who may find themselves in positions of high authority.

Extension services should form the bedrock of efforts by the governments of developing countries to encourage small-scale agricultural production. This will not undermine their efforts at import-substitution and quest for foreign exchange with current emphasis on plantations, importation of heavy agricultural machinery and distribution of agrochemicals. It is curious that while most of these governments agree that increased national food output can be easier achieved by stimulating small-scale production, policy conception and implementation clearly favour export-oriented large-scale producers. These entities can afford to pay for specialized and/or consultancy services without government assistance.

Adequate and sustained attention should be given to the human-resource component of the extension programme in order to remove or at least minimize constraints to optimal performance. Such limiting factors include insufficient staffing in terms of number and quality, poor housing and transportation facilities, poor remuneration, late or nonpayment of entitlements and lack of adequate incentives. The creation of a favourable working environment for field workers usually results in improved performance.

The effectiveness of extension programmes to a large extent depends on the existence of adequate physical infrastructure (including those dedicated to weather and vector-monitoring as well as communications). Timely and adequate provisions of subsidized inputs and support-services are also invaluable. Relevant support-services include plant and equipment hire, scientific research and health-care facilities for the farmer's families and their livestock.

Since governments usually provide support services, it is their obligation to assign roles and define jurisdictions for agencies directly or indirectly involved in agricultural extension. Government is also in the best position to ensure acceptable levels of cooperation between state-run agencies and other service-providers to ensure an optimal use of available resources.

Training programmes for extension workers should adequately prepare them for the tasks involved. In addition to a functional knowledge of agricultural processes, they should have a general knowledge of sociology, climatology, health-science, and soil and water conservation. At the intermediate levels, extension workers should possess *hands-on* knowledge and master the use of information and communications technology devices such as terrestrial and satellite terminals, computers and smart-phones. They should be able to access, analyze, collate and disseminate information relevant to their particular areas of operation. The employment of Drone Technology will lessen the need for the manpower and other rsources required for the survey and monitoring of extensive land masses. Many developing countries deploy the least qualified personnel to serve as field agents. This practice should be discouraged.

To supplement government efforts, rural farmers should initiate group action to improve their lot. The duties of extension workers would be much easier if they work with organized groups of farmers. An extension programme also has a greater chance of success where there is a history of successful group action within the community. Farmer-groups are also in a position to send their representatives to liaise with government agencies under whose jurisdiction they fall. Even where such agencies are unable to assign personnel to assist them directly, the agencies would be able to provide group representatives with useful information and educational materials relevant to their needs.

As a result of increasing variations in weather-trends, debilitating factors such as increasing soil-nutrient depletion and incidence of new pests and diseases, small farmers are finding it increasingly difficult to obtain acceptable yields with the farming methods which they are accustomed to. There is a paramount need for the farmers to receive assistance simultaneously from a number of facilitating agencies, if they are not to be discouraged.

Consequent on the above contention, state-run, well-conceived, implemented and funded agricultural extension programmes stand in exceptional positions to coordinate the various efforts aimed at assisting the small farmers to stay in profitable production.

Chapter Thirteen:

TROPICAL SOIL MANAGEMENT

Description

The term *soil* is usually defined from the standpoint of the discipline concerned. Soils are generally as small as, or smaller than grains in size. Geologists, archeologists and geophysicists all have their different soil definitions and classifications subject to the bias, content and demands of their profession. Since our concern with soil is limited to the agricultural perspective, we will adopt the definition most suitable for our purposes. We will therefore define soil as *loose earth surface formations resulting from combinations of minerals, organic matter, water and gases.*

Soils can be classified on the bases of use, altitude, chemical composition, hydrogen-ion content, drainage, organic-matter content, structure, texture etc. We will however adopt the system of classification according to texture since it is sufficient for our purposes.

Soil texture refers to the size of the individual particles of the soil. Under this method of classification, soils are broadly divided into three groups namely; sand, loam and clay. The size of the soil particles decrease progressively from sand to clay. A particular soil is termed sandy or clay depending on the dominance of the relevant soil particle size. Sandy soils have a rough or gritty feel while clay soils have a smooth or fine feel. Loamy soils consist of a mixture of sandy and clay soils. Moving from sandy to loamy to clay, soil becomes heavier with a greater capacity for water-retention.

Soil functions include anchorage for plants and acting as media for germination, gaseous exchange and the balance of soil nutrients. Human beings therefore depend on the soil for the satisfaction of their basic food needs. Man either consumes plant products directly rooted in the soil or products from animals which are ultimately sustained by plants, no matter their relative positions in the food-chain. Water-based plants and animals are also components of human diet but their quantities are however small in comparison with soil-based food items. This is especially true of developing countries with low per capita protein intakes. In this category of countries, aquatic animal products such as fish and crustaceans (crabs, prawns, shrimps, lobsters, etc.) are consumed in minute quantities and are strictly used as food seasoning or protein complements to plant-produce dominated diets.

One fundamental reason for soil being a necessary medium for man's existence is the fact that some essential compounds required for the synthesis of plant-food occur naturally in the soil. Growing plants continually remove these nutrients from the soil. This removal results in the depletion of the soil stock of nutrients, making the soil less productive or infertile. In a virgin soil that is not yet significantly affected by human activity, there is usually a balance between the loss of soil nutrients and its replenishment. Nature has its own peculiar ways of replacing lost soil nutrients. This natural replenishment process however takes place rather slowly when the soil is largely left unaffected by human activity.

Abuse

. As a result of man's interference with and disruption of the natural state of virgin soils, the fragile equilibrium where nutrient removal roughly equals nutrient replacement over time, is disrupted. Man alters the natural state of the soil through carelessness and/or in a bid to produce food and fibre for domestic and industrial uses which also leave naturally-stable soils open to harsh weather elements such as wind, rain and the sun's rays, accelerating the onset of erosion.

Generally, with prolonged mismanagement or abuse, it becomes increasingly difficult for soils to sustain profitable agricultural production. In addition, it has been alleged that soils in some humid and sub-humid regions of the world such as West Africa are naturally prone to crusting and susceptible to accelerated erosion when man tampers with its natural state of equilibrium with the environment.

Some traditional crop production practices contain elements of soil misuse or abuse, the most deleterious of which is the practice of bush burning. Bush burning is favoured in traditional agriculture as a convenient means of bush clearing. Bush burning is extremely harmful to the environment. While wood ash which results from bush burning undeniably adds some nutrients to the soil, the quantity made available is negligible in comparison to what raw plant tissues would add to the soil if incorporated through the natural process of decay. Wood ash is also highly prone to *leaching*. Leaching occurs when soil-water carries away nutrients dissolved in it. Bush burning exposes soils to wind and water which find it easier to remove nutrient-rich topsoil, than from soils covered with vegetation or debris. Bush burning results in the destruction of helpful live organisms resident in the soil and predisposes the soil to baking by the sun's hot rays.

Another form of soil abuse in traditional agriculture is the indiscriminate construction of mounds and ridges. Mounds and ridges which are the most common forms of seedbed preparation in traditional crop farming are sometimes constructed with little regard for the topography or slope of the land. Badly constructed mounds and ridges pave the way for accelerated soil erosion through the increased speed of water runoff after heavy rainfall or flooding. In addition, deforestation of hilly terrains leads to landslides during heavy precipitation.

Traditional crop farming encourages soils to be tilled or broken up every farming season so as to loosen its compactness and increase the number of air and water pores. With prolonged tilling however, there is a breakdown of soil structure. Soil structure implies the manner in which the component particles of soil come together or aggregate in order to form units. If the soil is too wet (as can obtain after prolonged or heavy showers of rain or in instances of over-flooding that sometimes occur in irrigated agriculture) tilling could also lead to a breakdown in soil structure.

Overgrazing and unregulated planting activities lead to depletion or over-exploitation of soil resources.

It should not be concluded that traditional crop farming is synonymous with soil abuse. Traditional crop farming makes some effort at soil conservation, albeit unwittingly. *Soil conservation* here connotes conscious efforts to preserve soil resources and fertility. Farm animals are driven onto fields after harvest, partly to consume crop residue and out of an awareness of the fact that animal droppings improve the soil. The cultivation of mixed crops even when carried out for economic or other reasons is an unconscious effort to balance nutrient uptake from the soil. Mixed cropping occurs in combinations among cereals, grain-legumes and root-crops grown on the same plot either at the same time or in rotation.

Soil abuse is not limited to traditional crop and animal farming. Modern methods of farming based on intensive agricultural practices contain inherent elements of soil abuse. The cultivation of sole crops which is a common feature of large-scale, modern agricultural production, overexposes soils to harsh weather elements. The mechanical process of soil preparation breaks up soil aggregates while the use of heavy machinery leads to solidity and the formation of hard-pans (soil compaction). The extensive use of agrochemicals also alters the form of fauna and microorganisms previously adapted to the soil and leaves harmful residue which may alter the soil's chemical composition and contaminate plant and animal products.

Management

There is increasing pressure on land available for crop production in the developing countries. This pressure arises largely from urbanization, industrialization, nature and wildlife conservation, tourism, boundary adjustments, breakdown of land-tenure systems as well as the exploitation of solid-mineral and other soil-based resources. Fertile, nutrient-rich wet-lands usually available to smallholders are being lost as a result of unrestricted land-reclamation and sand-mining activities. Agricultural land may consequently not be in such abundance as to sustain soil conservation methods like fallow or resting periods for depleted soils under cultivation. More efficient methods of soil management are therefore required.

Soil Management in agriculture essentially involves the maintenance or enhancement of a soil's state of profitable agricultural production without depleting its resources or causing physical harm to it. Soil management encompasses such practices that will conserve or restore soil fertility within the shortest possible time-frame. These practices should ideally employ the twin strategies of Soil Improvement and Soil Maintenance.

Soil Improvement is required when a soil dedicated to agricultural production is unable to carry out its basic functions. Even in the absence of diseases, pests and adverse weather conditions, agricultural plants may fail to perform satisfactorily in poor soils. Soil improvement entails such practices as Fertilizer-application, Flushing, Irrigation and Drainage.

Fertilizers are added to deficient soils to make them more productive. Fertilizers are basically grouped into two categories; Inorganic fertilizers, which are chemical compounds usually synthesized from Nitrogen (N), Potassium (P) and Phosphorus (K). The second category known as Organic fertilizers consist of natural, biodegradable matter.

Inorganic fertilizers are available either in the form of single compounds such as Urea or Phosphates or as mixed compounds (N-P-K). The type and quantity of inorganic fertilizers applied to soils depend on the nutrient requirements, soil peculiarities and type of plant to be cultivated. Effective fertilizer application requires the help of professionals to carry out soil and crop analyses to determine the type, quantity and method of application.

Organic matter consists of dead and decaying plant and animal tissues. Organic matter is a basic component of the soil and plays a significant role in soil improvement by adding nutrients essential for plant growth. Organic fertilizers also known as farmyard manure, are applied to soils in such forms as animal droppings (including wastes from poultry and other animal-farms), cut vegetation and weeds, wood and crop residue. Organic fertilizers serve as cementing agents for soil aggregates. Soil aggregation increases the air and water spaces in the soil, limiting the suffocation of agricultural plants due to water-logging or wilting due to excessive drainage.

Grouped as organic matter are live, active organisms such as insects, rodents, earthworms and other soil-burrowing animals. These organisms improve soil conditions by creating more air and water spaces as well as incorporating dead tissues into the soil. Not to be left out are the *decomposers*, which are soil microorganisms that initiate and complete the breakdown of dead plant and animal tissues.

Irrigation involves the controlled flooding of soils in order to increase the water-content of dry soils, making it more amenable to crop production. It should be noted that where the rate of underground seepage is inadequate, irrigation brings about problems of water-logging and excess mineral buildup, which cause further soil deterioration.

Drainage on the other hand, removes excess water contained in water-logged soils.

Where a particular soil is found to contain too much salt, improvement can be made to it by *flushing,* that is, irrigation followed by immediate drainage. Irrigation and Drainage obviously involves the use of tools and equipment. Technical know-how is also necessary to carry out these complex processes. Intervention by government, other development institutions or at the very least, Group Action is required by small farmers in developing countries in order to overcome problems of dry, waterlogged or acidic soils.

Soil Maintenance is carried out in order to check the depletion of a soil's productive capacity by replenishing nutrients continuously used up by growing plants. With high intensity of use and without proper maintenance, gains accruing from soil improvement can be easily lost.

An important aspect of soil maintenance is erosion-control. Soil erosion occurs when the nutrient-rich topsoil is continuously washed away by water or blown away by the wind. Water causes soil erosion mainly in flood plains as well as in the high rainfall zones of the world. Water erosion is minimized by keeping soils covered with vegetation for as long as possible, making ridges and building terraces across sloppy landscapes, constructing embankments, trenches and infield water runoff channels. *Strip-farming* which involves alternating cultivated plots with strips of grassland can also be embarked upon as a soil-erosion control measure.

Wind causes soil erosion predominantly in arid or semi-arid areas with low rainfall and high evaporative stress.

Wind erosion is controlled by planting tree lines or belts in a windward direction to serve as breaks. It is also controlled by keeping soils covered with vegetation for as long as possible.

Soil Maintenance also involves conscious efforts and measures aimed at replenishing nutrients lost through leaching. *Leaching* occurs when essential plant nutrients are dissolved in soil-water ending up in nearby rivers and streams. Soil Maintenance ensures that a correct and optimal balance of soil nutrients is constantly achieved. This process is implemented by controlling water run-offs and by the replacement of lost nutrients with elements from natural and artificial fertilizer sources.

As earlier noted, an uncultivated soil is more or less in a state of equilibrium with its physical environment. Man's exploitative and pollution-causing activities, in conjunction with climate change largely exacerbate problems such as erosion which occur when he tampers with the soil environment (fauna and flora), composition and structure. The proper selection and timing of farm activities in order to cause minimum disturbance to the soil's natural state serve as the bedrock of soil maintenance.

Soil-water conservation is an invaluable aspect of soil maintenance. Most soils can be conditioned to the cultivation of target crops. Conscious efforts must however be made to remain within the range of ideal-crop choices. *Ideal-crop* choices are those whose cumulative water uptake from the soil can easily be replaced by natural and/or available artificial means. The use of ideal crops ensures an optimal balance in the use, storage and replenishment of soil-water.

Special Soils

The ideal soil texture for most cultivated crops is loamy soil. Loamy soils consist of mixtures of sand and clay particles in approximately equal proportions. Agricultural soil available to a farming community may however not strictly fall within the category of loamy soils. Some crop species may also require a soil texture which is not loam for optimum performance. The cultivation of paddy-rice, for instance requires *puddling* which essentially involves the breaking-up of soil aggregates. Here, flooded soils are tilled in order to prevent water from draining away. Special treatment is required if special soils are to be useful for cultivating other species/types of plants.

Sandy soils which are characterized by the looseness of individual particles usually need improvement in order to make them subject to a wider variety of agricultural uses. The use of plant mulch (leaving plant residue on the soil after harvest) reduces overheating of sandy soils by the sun. Organic matter can be added to sandy soils in order to act as a binding agent, facilitating soil aggregation. It increases soil-nutrient status and water-retention capacity as well as encourage earthworm, microbial and insect activity which enables optimal plant performance.

Clay soils fall within the category of soils needing special attention in order to maintain its productive capacity. Proper timing of soil-preparation activities is essential for clay soils. Clay soils should not be tilled under very wet conditions since it hardens to become compact on drying. On the other hand, when clay soils are tilled under very dry conditions, it breaks up into heavy clods which pose difficulties during planting operations. Addition of organic matter and the cultivation of grass-forming crops such as rice, sorghum and millet which are integrated into the soil after harvest help consolidate and improve clay soils.

Cultivated plants do not perform well in acidic soils because their roots are unable to penetrate very acidic regions of the soil. *Liming* (which entails the addition of calcium carbonate to soils) is used to lower soil acidity and increase soil fertility. This option may not be feasible to the tropical small-scale farmer unless he receives technical and financial assistance from relevant state or other development agencies.

Innovations

Recent methods of soil conservation are derived from such established practices as incorporation of organic matter, cropping systems, land-use, tillage systems and soil-cover.

Intensity of land preparation forms the basis for a recent method of soil conservation. Here, initial land-preparation for virgin soils involves the mechanical uprooting or felling of tree with minimum disturbance to the topsoil. For savannah or shrub lands largely devoid of trees, the vegetation is simply slashed before planting commences. Cut vegetation is allowed to remain where they fall in order to serve as plant-mulch and as sources of organic matter. Planting is carried out without the use of ridges or mounds. Where plowing or other farm activities requiring the use of heavy machinery must be done, these are kept to the barest minimum. This method of land preparation reduces incidence of soil erosion and soil compaction. Its main disadvantage is that there is extensive use of agrochemicals since weed-control becomes a significant problem.

Another modern method of soil conservation is based on the cropping system. Mixed cropping is favoured over the cultivation of sole crops. Mixed cropping facilitates a balancing of nutrient uptake from the soil since the several plant species cultivated are selected based on the fact that they individually utilize more of different elements. This method ensures a fairly-even removal of soil nutrients and prevents a situation where plant growth is inhibited due to an excess or lack of specific elements. The down-side to mixed cropping is that mechanization of such activities as planting and harvesting become extremely difficult even with very simple implements.

Mono-cropping engenders the removal of specific essential plant nutrients such as Nitrogen from the soil in significant quantities. Some plants are however known to *fix* or introduce certain elements to the soil. Plants belonging to the legume family such as cowpeas are known to fix Nitrogen. One modern method of soil conservation is to *inter-crop* grain legumes with cereals and tubers in order to replenish the soil stock of nitrogen. An integral advantage in inter-cropping is that the longer the period soils are under plant-cover, the less vulnerable they are to wind and water erosion. Main crops may be planted in relay with edible or inedible legumes (or where possible in the off-season so that the soil remains under plant cover for most of the year). Where non-edible legumes are used, they could either serve as animal fodder or be plowed into the soil, prior to planting of the main crop.

Crop residue may be left on the field after harvest rather than burning them. The residue could act as fodder for farm animals which in turn enrich the soil with their droppings. Crop residue could also act as plant-mulch. Mulching, apart from integrating organic matter into the soil, prevents overheating. It also increases the capacity of the soil to retain water and reduces splash erosion, which is caused by heavy raindrops. Limitations to the use of crop residue as plant mulch include the possibility of greater pest infestation for future crops. Residue from diseased crops could also cause great damage to subsequent crops if used as mulch.

Live mulch has been used in a recent method of soil conservation. Here the main crop is planted in the midst of another plant species which should possess the capabilities of providing soil cover and soil enrichment. Legume plants appear most suitable for use as live mulch. Live mulch also has the added advantage of aiding weed-suppression. The main constraint to the use of live mulch is that it can only be used where water is not a limiting factor such as in flood-plains or in irrigated agriculture, since it obviously competes with the main crop for moisture.

No matter how physically or chemically balanced a soil may be, it could be unsuitable for profitable production if it is waterlogged or constantly flooded. This unsuitability also prevails if its water supply is inadequate. It is necessary to ensure the availability and optimal use of water available to small-scale agriculture in the face of increasing droughts, depletion of underground water (aquifers), decreasing water-levels in reservoirs, canals and dams, diversion of surface water as well as soil-drainage to facilitate civil construction works.

Chapter Fourteen:

STORAGE IN SMALLHOLDER AGRICULTURE

Losses

A sizeable portion of the agricultural output of many developing countries is lost due to inadequate preservation and storage techniques. Loss in our perspective implies depreciation in the nutritional or economic value of agricultural products or produce. It is avowed within some academic communities that some net food-importing developing countries, could conveniently feed their teeming population even without significantly increasing agricultural output, if they could drastically reduce their post-harvest losses. For our purposes, *post-harvest* period will be taken as commencing from the instant the animal or plant species is physically separated from its growth or production medium. *Growth or production media* refer to fields, sheds, pens or farmhouses depending on the species under consideration. Post-harvest period is assumed to end when the produce is consumed or utilized as inputs for other production processes.

Factors responsible for the post-harvest loss of agricultural products can be conveniently classified into four groups; Physical, Biological, Technical and Chemical.

Physical factors include; humidity, which is in turn comprised of moisture and temperature. Moisture means *degree of wetness* and is made up of two aspects in respect of produce; the degree of wetness of the produce itself and the degree of wetness of the air in the environment of the produce. Both aspects are interdependent. There are critical moisture and temperature levels beyond which agricultural products cannot store well after harvest. Spoilage organisms and pests of agricultural products are most comfortable and multiply rapidly beyond these critical levels. Freshly harvested crops being biologically active still respire. This respiration causes the produce to give up waste-products, heat and water at rates determined by the levels of temperature and moisture within the storage environment.

Biological factors include microorganisms, insects, birds and rodents. These organisms attack agricultural products by burrowing into them, eating them up, depositing toxins and metabolites in them as well as initiating discolouring of the produce. They also contaminate the produce with their dead bodies, skins, hairs, feathers and wastes such as urine and feacal matter.

Technical factors that initiate or predispose agricultural products to post-harvest losses revolve around faulty handling methods. Produce may be harvested at the wrong time or using flawed techniques, while defective methods of drying, transportation and storage may be employed. The produce may also suffer much damage during such initial processing activities as cleaning, shelling, hulling and threshing.

Chemical factors affecting post-harvest losses include the degradation of agricultural products as a result of intrinsic biochemical (aging) factors. Also, during pest control and storage programmes, chemicals may also be applied to the produce. These chemicals accelerate the rate of deterioration of harvested products where it is not judiciously applied.

Physical as well as biochemical damage sustained

by produce increases its susceptibility to further deterioration or attack by spoilage organisms.

Crop farmers face severe problems of loss of produce even before any storage activities can take place. Pre-storage crop losses take place in a number of ways; deterioration of the produce begins in the field as soon as the crop matures prior to harvest, losses during harvest and initial processing activities, losses which occur as the produce is being moved to assembly points, processing or storage facilities. Finally, losses occur when the produce is placed in temporary storage or holding-facilities.

Crop losses can be categorized as Weight-loss, Food-loss, Quality-reduction and Seed-loss. Weight-loss results when attacking organisms eat up part of the produce while Food-loss occurs when the nutrient-content of the produce is affected as a result of the feeding activities of organisms. The protein, starch or other biochemical constituents of the produce could be altered as a result. Quality reduction occurs when attacking organisms discolour, leave holes, wastes, eggs or larvae in the produce. Finally, Seed-loss occurs when organisms inflict such damage to the produce that it loses its viability or reproductive capacity. When this happens, the seeds are no longer useful as seedlings for planting purposes.

All the losses mentioned above translate to Economic loss for the farmer since he may end up selling lower volumes, lose trading opportunities or have to sell at a discount.

Preservation

Ideally, a small farmer's scale of operations is expected to increase in size with time. This increase is facilitated by the injection of fresh capital, use of improved seed-stock and employing more efficient systems of cultivation. This increase in output creates additional problems for the farmer if he does not dispose of all of his products as soon as it is harvested. The farmer may not be able to, nor have the inclination to dispose of all of his produce at harvest time. The farmer may wish to retain some of his produce for seed or consumption purposes. Local markets may be saturated or prices too low. These possible scenarios would accentuate the problems associated with a lack of adequate arrangements for produce-preservation by the smallholder.

Some preservation and storage activities take place even in peasant agriculture where the levels of output are consistent with production circumstances. Yams are tied to staves in barns, cassava is left in the ground upon maturity until needed and grains for domestic use are usually sun-dried or processed into more durable forms such as flour. Seed-grain for crops such as maize is hung by the fireplace. The firewood, besides its drying effect, emits smoke particles containing compounds with preservatives. These preservation and storage techniques though having served the peasant farmer's needs over a long period of time, become inadequate for his purposes as his scale of production increases or as he evolves to more modern forms of agriculture.

If preservation and storage technologies were simple, inexpensive and could be universally put to use, small farmers would not have to dispose of their products immediately after harvest or at unfavourable prices. There is also a high level of probability that they would strive to expand output in order to reap more economic benefits.

Most modern post-harvest storage systems are usually produced through location-specific problem identification and specialized technology based on optimal solutions. That is, the knowledge about available storage systems, local pests responsible for post-harvest losses as well as climatic peculiarities is used to devise optimal solutions for each locality. Extension personnel, other agents of government, educational and agricultural research institutions, corporate concerns and private individuals work in concert to provide these solutions.

Three approaches are generally adopted for facilitating the effective and efficient storage of products in small-scale agriculture. These are; employing new varieties of plants less susceptible to post-harvest storage losses, improving on-farm storage and centralizing storage especially in the case of cereals.

Levels of Storage

Post-harvest storage of produce can take place at the field-level. Field storage is mostly emergency storage. It is usually employed when immediate disposal is anticipated or as a temporary measure pending evacuation of produce to safer storage facilities. In the case of cereals, field storage may be carried out to allow grains to become dry enough for alternative forms of storage. Field storage also allows some initial processing activities such as shelling, winnowing and threshing to be conveniently carried out.

In field storage, bunches of paddy-rice could be hung inverted in the field to dry. Un-threshed grain could be tied together and left erect on their stalks or set on crossbars in such a way that allows breeze to blow through it. Harvested grain could either be piled on wooden planks or any other available materials such as banana leaves or plastic sheets. It should be noted that this type of storage tend to leave the produce at the mercy of rodents, birds and insect pests.

One relatively safe form of field storage entails the use of cribs. *Cribs* which are elevated rectangular cages could be of any length but of restricted width. A crib essentially consists of a raised platform which is supported on wooden, bamboo or metal limbs. It has its sides covered with wire-mesh or some other material which allows unrestricted ventilation while at the same time limiting access to rodents, birds and bigger pests. It is usually roofed with grass, raffia or metal sheets. Cribs are used to store de-husked maize, panicles of paddy-rice, sorghum and unshelled cowpeas. To check possible infestation by flying insects, the grains may be dusted or spayed with recommended chemicals. If dusted or sprayed with chemicals, the grains are not to be consumed by human beings or domestic animals until the recommended period of time has elapsed since the chemicals may be toxic.

Storage of agricultural products can be carried out at the farmyard level. Farmyard storage reduces transportation costs to a minimum. There is also greater security for the stored products than in the case of field storage. An added advantage is that cereal grains can be easily re-dried when necessary in a bid to control its moisture-content. Pest control however poses a significant problem in farmyard storage. Root crops like cassava suffer heavy post-harvest losses when stored for more than a few days. They are consequently left in the ground until needed. They can however be preserved for longer periods after harvest if they are stored in cartons, boxes or bins. An adaptation to the ordinary container storage in the case of root crops is that harvested roots can be placed in moist sawdust or debris from deep-litter poultry sheds. This way the roots can have greater longevity extending to a couple of weeks or more.

Another farmyard level storage technique is the semi-underground storage facility. This can be used to store sizeable amounts of produce. This facility is usually employed in arid, semi-arid regions or places where the soil is not prone to water-logging. In semi-underground storage, a small pit or trench is dug with its sides raised above ground-level. The floor and sides of the pit is plastered with cement or mortar. Dried, shelled or hulled cereals or pulses are poured into the pit, sprayed or dusted with recommended chemicals, covered with planks or leaves and finally, a layer of soil. Root crops can also be kept in semi-underground storage pits in which case palm fronds or some other similar material are used as intervening layers between the layers of roots, before sealing with earth. Storage pits or trenches should be opened at reasonable intervals to check for signs of produce-deterioration or pest infestation.

Storage at the farmyard level can also take place in

sheds, barns and huts. Some tubers can be stored for up to six months in well-ventilated and protected barns. Since tubers are physiologically active, natural changes in weight and quality cannot be entirely prevented. In order to achieve the best results, tubers free from damage or bruises should be staked on open-sided shelves constructed with live wooden poles, bamboo or sawn wood. Buds should be removed as soon as they appear, while rotting can be controlled by the use of appropriate fungicides. If tubers are to be stored upright, it must be placed on their heads.

Grains could be stored in circular brick or mud-walled huts roofed with thatch or some other more-durable material which should be a poor conductor of heat. Shelled, de-husked or hulled grain can also be stored at the farmyard level in completely filled, airtight containers such as plastic or metal drums. It should be stored at safe moisture content (i.e. not too dry or too wet). Polythene bags may also be used for grain storage where there is adequate protection against attack by rodents. The grains should ideally be sprayed with pesticides before being sealed. The pesticide may be in tablet-form in which case it should be placed in the container with the produce before sealing takes place.

It is very important to note that grains near to consumption or sale should not be treated with chemicals. Application of chemicals to agricultural products meant for consumption must be carried out under the direction or supervision of trained personnel since they can be extremely harmful if not carried out in the prescribed manner.

Large-Scale Grain Storage

One of the solutions to problems of preservation and storage in small-scale agriculture is the centralization of storage. Centralization of storage is easier in the case of grains belonging to the legume and cereal families. These grains include maize, sorghum, paddy-rice, millet and cowpeas. Centralization is dependent on the availability of medium to large-scale storage facilities which include grain-silos, warehouses and stores. The provision of modern storage facilities is beyond the financial capability of the typical small-scale agricultural producer. Interventions by governments or other development agencies are required if large-sized facilities are to be put in place for use by small farmers.

Centralization of storage means that some degree of cooperation has to take place among the small farmers since it necessitates the efficient collection, drying and/or initial processing of the produce. There are some issues to be ironed out by the farmers concerning grading of the produce, cost of processing and storage, disposal of the produce and sharing of the proceeds. It might make better sense for small farmers working on commonly-held farms to procure and make use of medium to large-scale facilities while individual small farmers limit themselves to storage at the farmyard level.

Grain *Silos* are usually operated by governments or big private concerns which procure them for the storage of their produce and/or produce procured from other producers. A Grain Silo is a medium to large-scale storage facility, consisting of a tall cylindrical structure usually assembled from prefabricated sections made from plastics, metal or moulded concrete. The silo design allows grain to be loaded at the top and removed from the bottom sections.

Warehouses can also be used for medium to large-scale grain storage. A warehouse specifically designed for grain storage essentially consists of a large well-secured, concrete building inaccessible to birds, animal and insect pests, with possibilities of controlled ventilation by the use of windows and electric fans. The roof and/or upper sections of grain warehouses ideally should have portions covered with transparent sheets to allow sunlight into the interior of the warehouse, while the bottom sections should have porous makeshift floors raised a little distance from the ground.

Grain stores are smaller, less complex variants of warehouses. A store must not as a rule have transparent windows, electric fans or a raised porous floor. Grains stored in bags should be placed on wooden pallets and not in direct contact with the concrete floor and walls when preserved in grain stores.

Ideal Conditions

Before long-term storage of grain takes place, allowances have to be made for the peculiarities of each of the species. For instance, sorghum is more susceptible to damage by insect pests than finger-millet. An essential prelude to good storage is to ensure that the moisture-content of grains is as near as possible to predetermined ideal values. *Moisture-content* refers to the amount of free-water contained in the grain. Free-water means the volume of water which the grain will give up upon drying. Cereal grains may be harvested at moisture-levels well above the safe range for good storage. Wet grain must however be subjected to drying in order to attain the ideal moisture content (which is about twelve to thirteen percent), before long-term storage can take place. This ideal moisture content allows grain to be stored alive so as to resist deterioration better and should be maintained throughout the storage period.

The usual method of reducing moisture content of very wet produce to acceptable levels for long-term storage is by drying. This is carried out naturally by exposing the produce to direct sunlight or artificially by allowing the passage of warm air through the produce, using mechanical heat sources or conventional electric heaters.

When drying produce naturally, the rate and extent of drying is dependent on weather conditions. Consequently drying is faster where the initial moisture content of the produce is high, humidity is low, temperature is high and airflow or wind activity is increased. The rate of natural drying is also dependent on the mode of harvest and post-harvest handling of the produce.

Natural drying of produce is cheap and convenient when compared with artificial drying methods. However, the non-predictability and non-reliability of weather elements, longer period for attaining optimum moisture levels, continued exposure of the produce and uneven drying are limitations to the use of the natural drying method. Artificial drying methods though not easy to implement have such advantages as;

enabling earlier harvesting of produce to reduce risks and decreasing the chances of deterioration while in storage.

As important as drying is to grain storage, it can lead to extensive losses if not carried out well. Uneven drying, which connotes the phenomenon where the outer parts of produce undergoing drying are brought down to required moisture levels whereas the inner parts still possess high levels of moisture, is undesirable for grains destined for long-term storage. Again, when grains are dried beyond the specified safe moisture content, they are rendered unfit as seedlings for planting and such industrial uses as malting and wet-milling.

Grain Store Management

Grain stores being smaller, less complex storage facilities are the most likely type to be used by small-scale farmers who individually or collectively may have been able to acquire them, having chosen the option of deferring the disposal of their produce. Heavy losses may however occur within grain stores if the facilities are not well constructed and managed.

Stored grain must be maintained at the recommended storage moisture-content to minimize the incidence of insect attack and the growth of mould. Grains should be kept cool during storage since high temperatures increase the rate of grain respiration, leading to further increases in temperature and moisture levels. Clean and if possible, chemically-treated sacks and containers should be used for storing grain. Regular checks of the stored grain (not beyond eight-week intervals) should be made in order to detect possible deterioration or pest infestation.

Insects are the most important and persistent attackers of stored grain. Infested grain may be spread under the sun to drive away insects. The main limitation to this strategy is that sunning of grain may not kill insect eggs and larvae. Some easily-removable, insect-repelling local herbs without residual tastes or odour may be mixed with stored grain where available. Airtight containers may also be used to check insect infestation since the insects would be unable to breathe. Contact chemical insecticides or repellents may be mixed with or sprayed on the grain as it goes into storage. The interior surfaces of containers and storehouses may be sprayed or brushed with insecticides, while fumigation may be used as a last resort to get rid of serious insect infestation. The grain store itself should be clean and its surroundings constantly free from debris and vegetation.

Grain store owners should generally keep an eye on advancements in anti-pest technology. Advances in solar and wind powered technology facilitate the availability and affordability of equipment necessary for keeping grain stores at ambient temperatures and humidity.

Some technologies based on *Electronic Pest Control* are universally available. These relatively cheap devices use radio-waves (ultrasound) to repel insects and rodents. These devices where considered for use in agricultural storage will eliminate problems of toxicity associated with chemical pest control. The down-side to the use of this device is that it will have to be constantly modified since the pests will after a period of time, adapt or develop levels of tolerance to the frequency employed.

Chapter Fifteen:

MARKETING IN SMALLHOLDER AGRICULTURE

Evolution

Agriculture is primarily responsible for the genesis and initial growth of the domestic economies of a significant number of tropical developing countries. Some notable events result in the accelerated growth phase of these countries. These events which include the exploitation of agricultural, human or mineral resources and some political exigencies, lead to urbanization, increased commerce and industrialization. Direct consequences of these events include increased rural-to-urban migration, industrial raw material needs and more urban needs for food and fibre.

Urbanization and rural-to-urban migration subsequently change the predominantly rural landscape occupied mainly by peasant and small farmers. Manpower is gradually displaced from peasant agriculture to other areas of gainful employment, lured by greater financial returns and less physically-stressful means of generating incomes. The declining numbers of small farmers left behind in rural localities are thus able to dispose of all of their farm products and are even hard-pressed to cultivate enough food, fibre and industrial inputs to meet the increasing domestic needs.

With time, dreams of finding fame and fortune in urban centres fade as people face the realities of insecurity of life and property, nonexistent or declining job opportunities and the high cost of living. Many people consequently return to small-scale farming. Some people previously engaged in other pursuits resort to agriculture as a means of livelihood, while others engage in *back-yard* farming in order to reduce their feeding expenses or supplement family incomes.

As domestic economies continue to expand, large-scale farming concerns begin to emerge. There are also sustained efforts by governments in struggling economies to encourage agricultural production in order to ensure food security, facilitate import-substitution and increase export as well as foreign currency earnings.

With the above occurrences, marketing begins to assume a more prominent position in the hierarchy of the small-scale farmer's problems, since there is increased domestic output of agricultural products to the point where possibilities of glut exist.

Marketing is a general term, which encompasses all activities geared towards the movement of commodities or services on offer to the public from its point of production to its point of consumption. By extension, agricultural marketing activities would connote the various tasks undertaken in order to move produce beyond the farm-gate to its final consumers. In addition to buying and reselling produce, these tasks include preservation, storage, transportation, processing and grading. Advertising is a component of marketing, but it is usually beyond the reach of the rural small farmer who remains our focal point.

Issues

The multifaceted problems associated with the disposal of agricultural products become more obvious to a small farmer who is not engaged in contract-farming. First, under rain-fed agricultural production, the farmer has to contend with seasonality. *Seasonality* is the phenomenon where there are alternating periods of wide spread scarcity and abundance of specific produce depending on the length of its production-cycle and storability. This results in periods of inflated and depressed produce-prices within the same calendar year.

The bulky but delicate nature of most agricultural products precludes easy transportion to locations where better prices could be obtained for it. Where the rural farmer is lucky enough to have his personal means of transportation, it is usually crude and of limited capacity. He also cannot cheaply move his produce by public transport to identified markets because his level of production is too low to take advantage of economies of scale.

From the time the produce is due for harvest, the farmer has to protect it from pests and shield it from deterioration so as to prevent its market-value from declining. In spite of these exertions, he has limited time within which to profitably dispose of his produce, since good preservation and storage practices can only maintain the quality of produce up to a certain time-span, after which he has to dump it on buyers, no matter the prevailing price or risk outright loss of revenue. Moreover, he probably needs the money to take care of his family's needs and other pressing obligations.

Prior to production, the commercial small farmer usually carries out market research however rudimentary, in order to ensure that a profitable market exists for his produce.

At the onset of harvest however, the farmer has to devote quality time to locating willing buyers for his produce if he does not want to be stuck with it or be taken advantage of. Necessary farm activities aimed at preserving the produce may be neglected or not carried out well during the farmer's absence from the farm. Even with capable hands, some very crucial decisions may have to be deferred till the farmer's return, resulting in delays and possible losses.

With this cursory examination of the post-harvest issues from the small-scale farmer's perspective, it becomes obvious that it would be more advantageous if the tasks associated with produce disposal are not exclusively carried out by the producer. Some other entities are needed to undertake some of the crucial tasks necessary for the exchange process between producers and final consumers.

Middlemen

Markets for agricultural products can be conveniently grouped into three categories; *Consumer, Industrial and Reseller* Markets. In Consumer Markets, buyers obtain produce for consumption. Industrial Markets are where individuals and groups purchase produce for use as inputs for other production processes. Reseller Markets are mainly composed of *Middlemen*. The Middlemen of agricultural marketing consist of individuals and groups who are involved in the transfer of produce from farmers to consumers, even though they are neither producers nor consumers themselves.

Three categories of Middlemen exist in agricultural marketing. The first category known as *Merchant Middlemen* concentrates on the physical activities associated with the commercial process of buying produce and selling it either on wholesale or retail bases. This category is the most visible since they have the greatest number and spread among the three categories. They undertake the responsibility of transferring produce from the farmer to the final consumer at a cost which is referred to as *Marketing Margin*. Marketing Margin is the compensation received by the middleman and comprises *rent* (returns to capital utilized), *interest* (returns to money tied down), *wages* (returns to time and effort) as well as *profits* (returns to risks borne). Merchant middlemen introduce efficiency to the marketing process by using the available processing, storage, transportation and sales facilities maximally and introducing new techniques and technologies as needed.

The Merchant middlemen of agricultural marketing assume the risks and uncertainties associated with agricultural marketing. Risks here include possible losses from inadequate storage, handling or transportation, while uncertainties include possibilities of adverse price and demand situations.

In situations of inadequate preservation technology and storage facilities, small farmers wish to dispose of most of their produce at harvest-time whereas domestic consumers wish to purchase varying quantities and grades as the need arise. Merchant middlemen undertake the arduous tasks of bulk acquisition and the sorting, grading, packaging and retailing of produce in smaller quantities over time.

Middlemen play other crucial roles in the marketing of

produce. The functions they routinely undertake include market finance. On the average, farmers want to receive cash for their produce right from the commencement of harvest. Final consumers on the other hand do not usually part with their money until the point of physical transfer of the traded produce to them. Merchant middlemen commit their financial resources towards holding the produce during the interval. Merchant middlemen have been known to get involved in the financing of production activities. They have provided credit to small farmers in such forms as cash, seed-stock, animal health-care products and agrochemicals. Others have teamed up with farmers to carry out contract farming.

The second category of middlemen known as *Brokers* serves as links between producers and consumers of agricultural products. This category does not commit its financial resources to acquire or hold the produce, but merely act as *Commission-Men* who bring buyers and sellers together for a fee.

Facilitating Middlemen, the third category of middlemen do not primarily undertake commercial activities but assist the marketing process. This group includes trade associations and cooperatives concerned with agriculture. This category of middlemen helps develop agriculture and broaden the market by guiding smallholders to produce according to specifications desired by importers and large-scale buyers of their produce.

The middlemen of agricultural marketing play a major role in market expansion as they constantly strive to improve sales by locating and harnessing potential demand opportunities. Their expansion activities widen the spatial dispersal of the produce ensuring its availability in places far from production sites, while preservation, partial processing and storage lengthen the time period during which seasonal crops can be enjoyed. *Produce-processing* carried out by some middlemen ensures that produce can be utilized in the most desirable forms. For instance, cassava roots can be processed into garri, fufu (meal), flour or industrial starch.

Middlemen unwittingly constitute a source of feedback to small farmers. Their levels of demand and the prices they are willing to pay for specific produce place farmers on the alert about the current state of the market, trends in consumer tastes and preferences. Production processes are thus deployed to the satisfaction of current demand, conserving scarce resources and reducing the level of wastage.

Theoretically, middlemen are expected to increase the levels of efficiency within the agricultural marketing system. First, marketing being their main preoccupation, they have the advantage of efficiencies arising from specialization. They also have the advantage of economies arising from operations in such tasks as bulk purchases, large-scale transportation as well as the optimum use of available labour, storage and processing facilities. This efficiency is supposed to impact on the final consumer in the form of reasonable produce-prices.

The merchant middlemen of agricultural marketing in most developing countries are thought to demand too much remuneration for their marketing activities, thus escalating the prices of produce to unreasonably high levels. The most significant factor affecting the final price per unit of produce is actually the length of the chain of intermediaries between the farmer and the final consumer. The more the number of middlemen for a particular produce, the higher its final price. The proportion of the final price per unit of produce received by the farmer consequently decreases as the number of middlemen involved in its transfer increases.

Small farmers can obtain a greater percentage of the price per unit of produce paid by final consumers, if they are willing and able to assume some of the functions performed by the merchant middlemen. Assuming marketing functions such as partial processing, preservation and storage which are intrinsic functions of the *Agricultural Value Chain* as well as greater involvement in transportation, grading, packaging and retailing would ensure greater returns for the small farmers. They can carry out these functions individually, as members of cooperative groups or marketing unions.

Intervention

The ideal situation for small farmers is to be able to operate at a profit and to obtain the maximum possible percentage of the price paid per unit of produce by final consumers. The ideal situation for final consumers is for them to obtain produce where and when needed, in the required quantities, grades and at the lowest possible prices. In developed economies, the interplay of market forces may be largely relied upon to stabilize the desires of both producers and consumers at an optimal level. However in the developing countries, economic parameters cannot be solely employed since too many extraneous factors are at play, causing distortions within the domestic produce markets. Interventions are necessary to ensure that undue advantages are not enjoyed by the small farmers, middlemen or final consumers.

The governments of developing countries are responsible for the promulgation of microeconomic policies which include the definitive role of protecting both producers and final consumers from adverse market situations. These adverse market situations include low producer returns, high consumer prices and significant price changes arising from seasonality of produce, scarcity of essential inputs, fluctuations in output volumes, climatic/weather peculiarities, depressed demand and excessive margins.

Bulk storage and irrigation facilities would help alleviate problems of seasonality. The provision of all-weather access roads to production centres coupled with an efficient public haulage system ensure easier and cheaper transportation of produce. The above measures in turn facilitate marketing activities.

Staple foodstuffs which small-scale producers are largely responsible for have been known to suffer severe seasonal fluctuations in output volumes. Governments at every level could maintain buffer stocks of these produce in order to ensure year-round availability and minimize price fluctuations. The procedure for this scheme entails the purchase of produce at reasonable prices to prevent a glut at harvest time. It is then sold to consumers at near purchase-prices when scarcity of the produce occurs in local markets. Strategic reserves for staple produce are beneficial to the market both from the producer and consumer points of view.

The necessary dissemination of invaluable market information to small farmers would be best handled by government since it is more likely to possess the required manpower and usually have considerable access to mass-communications media. Relevant information here includes that which will assist the small farmers to locate markets, ascertain current prices and determine consumer preferences. This information enables the farmers to fully exploit existing marketing opportunities.

Dismantling cross-border restrictions on the movement of agricultural products facilitate marketing and sales. However, there has to be some form of produce-inspection in order to prevent the spread of pests and diseases. Cross-border trade may also have undesirable consequences where the farmers across the border enjoy technology advantages or substantial subsidies which enable them to produce at lower cost.

Government-sponsored agricultural research in such areas as food preservation, processing and storage leads to an expansion in demand since it increases the period within which some seasonal produce can be enjoyed and makes available a wider variety from which consumers can make choices. Research should also be expanded to domesticate, improve and reintroduce neglected *wild foods* which used to contribute to the nutritional regime available to indigenous communities in the not too distant past which would excite consumers.

Government is in a position to facilitate the financing of agricultural marketing activities by formulating policies enabling easier access to short-term credit by farmers and middlemen. Formal insurance agencies could also be encouraged to get involved in agricultural marketing through mass enlightenment and by liberalizing the issuance of policies in respect of loss or damage to stored produce and goods-in-transit to groups of small farmers or marketing associations.

In most developing countries, policies which give rise to increases in real disposable income such as creating more employment opportunities and wage increases translate to increased demand for agricultural products. This is because the food component of family expenditure constitutes a significant proportion of average family income. The enactment and pursuit of policies aimed at wealth-creation and financial empowerment of the masses should form an integral part of national planning for economic development. This is even more important in the present era because as world population ages, a declining proportion of younger people have to provide for an increasing population of the older generation.

At the local level, governments of developing countries could take advantage of an oft-neglected means of generating additional revenue with which to fund interventions in small-scale agribusiness development. Self-employed persons and private-sector employees of small production and service ventures hardly pay tax on revenue earned, Working within the confines of national laws governing taxation, local authorities must find the political will to enact and enforce efficient and equitable sales tax schemes as well as personal income tax collection processes.

Chapter Sixteen:

CREATING SELF-SUSTAINING SMALLHOLDERS

Rationale

A significant number of the Tropical and hotter Sub-Tropical countries cannot be said to enjoy relative economic and political stability. The increasing radicalism of faith-based, ethnic and other social groups, clamour for good governance, more equitable distribution of commonwealth as well as battles for the control of mineral and natural resources are ostensibly major factors contributing to this instability. The underlying factors fuelling these social upheavals and discontentment are usually *Stagflation* (stagnant economies coupled with high inflation-rates) or hyperinflation, insufficient welfare programmes and lack of buffers against natural and other disasters as well as unacceptable levels of unemployment.

The typical scenario within the above category of countries is a situation where there is continuous official and unofficial devaluation of the local currency, with the economy not creating enough jobs to keep up with an increasing urban population and adult work-force. To worsen the situation, declining capacity occurs as a result of de-industrialization largely caused by; inadequate security of lives and property, insufficient raw materials and power supply, infrastructure decay and an apparent inability to compete with cheaper legally imported and smuggled goods, frequent dislocations as a result of man-made or natural hazards in addition to economic mismanagement.

Curiously, the *Food-Price* component is a significant constituent of the *Consumer Price Index* (a measure of the rate of inflation) within the aforementioned countries. It stands to reason that manipulating agricultural production would have a profound effect on prevailing inflation rates. It is also worthy of mention that their economies tend to possess largely underdeveloped agricultural sectors as well as under-utilized agricultural resources with vast potential for employment generation and wealth creation. The rates of inflation in these countries can obviously be kept within acceptable limits by having more agricultural products per unit of local currency, consequently cooling down their overheating economies.

Governments of the countries in view usually embrace and encourage large-scale farming in their efforts to reduce food and raw materials import bills, ensure food security, stabilize local food prices and earn foreign currency. These large farms normally rely on heavy industrial equipment and automated processes. Without dispute, large-scale farming strives to achieve greater efficiency by substituting human and animal muscle power with machinery and automation in the performance of routine tasks. Large-scale farming places little reliance on personal skills and initiatives which does not allow for the full exploitation of farm-family potentials.

Though able to reduce scarcity by increasing the quantity of agricultural output per operator, large-scale farming is not very effective at reducing unemployment, which is a major problem for the tropical developing countries. Farmers producing on a small scale can achieve a cumulative level of output comparable to that obtained through large-scale farming, with the added advantage of utilizing higher levels of manpower and low-skill labour, thus reducing the level of national unemployment.

Large-scale farming has some drastic consequences such as changing the composition of local fauna and flora as well as reducing bio-diversity. It also encourages large-scale deforestation and land-grabbing by speculators. The extensive use of industrial chemicals (especially pesticides) has been linked to debilitating illnesses among people living within the vicinity of large farming concerns.

Policy-makers in tropical developing countries may choose to tackle unemployment as well as achieve other macroeconomic goals by utilizing the opportunities presented by agriculture. They have sufficient room within their economies for both large and small-scale farmers to thrive side-by-side. Policy-makers should maintain or even increase efforts aimed at facilitating increased production by farmers currently in production as well as encourage people who have abandoned agriculture for other occupations to return.

In addition to the above measures, the seed for a new generation of self-sustaining smallholders should be planted through the creation of new carefully-conceived farms. Other advantages derivable from the creation of new smallholders include, but are not limited to; decongesting urban centres, facilitating rural development, reducing rural-urban migration, stimulating demand through economic empowerment, imparting new and useful agricultural skills, creating entrepreneurs who are future employers of labour as well as encouraging *Agribusiness* (production, storage, processing and distribution of farm-products and items made from them) thus developing the Agriculture Value Chain.

The creation of new smallholders would best be handled by authorities at the federal (national) and state (provincial) levels. This is mainly because of the magnitude of resources needed and new legislation that would be required to initiate and sustain the process.

The Process

TARGET PRODUCTS

The process of creating new smallholders should commence with the identification of staple agricultural products whose output can be substantially increased without creating a glut and depressing prices. Priority must therefore be given to; products with export potential, products used for import-substitution, products which can be processed into other varieties to stimulate demand and increase its longevity (post-harvest life), products serving as feed-stock for industrial production, products which investor-friendly incentives can attract entities operating within the food, beverage, bio-fuels, textile and leather products sub-sectors to set up local processing plants. Some categories of produce used in strategic food storage programmes aimed at enhancing National Food Security would also be ideal for this scheme.

LAND AND LEGISLATION

The greatest obstacle to the creation of new smallholders is access to land for farm-related activities. Tracts of unexploited and under-utilized arable land need to be identified and acquired. Abandoned and unprofitable public and privately-owned agricultural settlements, plantations and large farms should also be slated for acquisition. Arbitration and adequate security arrangements could free up lands in dispute and lands under constant attack by invaders.

New arable lands could also be created through restoration of lands unfit for profitable agricultural production (due to overuse, over-acidity or industrial pollution, deforestation, canalization, damming and irrigation).

Taking possession of large tracts of virgin land, existing farms, plantations and ranches would as of necessity require a legal framework involving legislation and arbitration, since issues concerning titles and ownership, boundary delineations and compensation may have to be dealt with. Legislation may also be necessary in order to create enabling structures, guarantee adequate and regular funding as well as ensure continuity of the scheme beyond the tenure of the founding government.

In order to maintain the ecological balance, relevant surveys and environmental impact assessments have to be carried out where extensive, contiguous land masses are to be put to use.

Safe-guards also have to be put in place to protect the schemes from the pervasive influence of criminal cartels, corrupt bureaucrats and unscrupulous politicians.

WORK-FORCE

A comprehensive list of unemployed/underemployed persons without gender restrictions, willing and able to take up farming as an occupation, should be compiled. These persons should state their preferences among production choices. They must be ready to relocate to new places of residence if necessary. They are also to be made to understand that farm-holdings are not being made available to them free of charge but on renewable leasehold bases.

Where acquired tracts of land are not easily accessible to urban centres, provisions have to be made for secure mini-residential estates equipped with basic amenities (power and communications) and have motorable access to primary healthcare and educational facilities where imperative. The smallholders on the other hand could be integrated into nearby population clusters. Conscious efforts should then be made to ensure that these locations form the nuclei of new modern villages/towns aimed at gradually decongesting overpopulated urban centres.

FARM SIZE

Taking into consideration product-type and site-location, the ideal farm-size would be based on the expected annual turnover of each holding. (Each farm is to be considered an independent business unit). The operating standard could be; *what farm-size would generate an annual margin (turnover less costs) equivalent to the National Minimum Wage X 3*. A fifty percent extra capacity should be added to take care of the future growth of each holding. This computation should not be a daunting task for agricultural-economists, who as of necessity should be assigned to the scheme.

Subsequent increases in the value of the holding are to depend on the level of skill and dedication of the smallholder. Product-price adjustments usually compensate for upward movements in cost of production due to inflation or other causes.

SITE PREPARATION

Depending on the product-type, site preparation would include land-clearing and/or drainage, construction of dams and water channels, access roads, boundary demarcations, sheds and barns for use in animal husbandry (poultry, piggery, etc)

For existing aging plantations and animal farms acquired for the scheme, the plants/animals may have to be replaced with higher-yield species. The primary objectives of site preparation are to shorten the time period required to achieve the initial harvest and to ensure that major equipment and services will not be required after hand-over to recipients.

SELECTION AND TRAINING

Using the database of potential smallholders, available farm holdings are to be allocated according to relevant experience, product preferences, location and other relevant criteria. Each cluster of new smallholders should ideally have at least one person with special training or prior farming experience whose farm will serve as a *demonstration or model-plot* for others, especially the new farmers.

Immediately after the selection process, workshops should be organized for the prospective smallholders to enlighten them on such issues as efficient production techniques, soil and water management, prevention and control of diseases and pests, on-farm processing and storage, supplementary farm activities, (snail-farming, fish-farming, etc), mixed farming, dealing with natural and man-made disasters, profitable use of idle-time and other relevant issues. By implication, this new generation of smallholders should preferably be literate, numerate and trainable, so as to increase the chances of survival and growth of their individual enterprise.

Finance and Implements

The new generation of smallholders is expected to be self-sustaining on the long run. Allowances must however be made to ensure their survival, at least to the first harvest and sales of their products, after which they may be left to fend for themselves. Guaranteed credit on liberal terms should be made available to those who need it. This credit should be in the form of provision of seed-stock, tools and implements, agro-services, cash and other inputs. The credit which is meant to cover production as well as living expenses should not exceed twenty-five percent of each smallholder's projected first year income. Smallholders may choose to access only production credit if they are able to survive on their own resources.

Support

Where there are significant numbers of smallholders within a production cluster, mini Support-Centres could be provided for them. These centres should be initially manned by small teams of agricultural support and extension workers who are trained to coordinate, facilitate and if necessary manage such activities as climate-watch and weather monitoring, implements hire (tractors, farm-tools and other accessories), procurement of farm tools, spares, improved seed-stock, agrochemicals and inorganic fertilizers. They should also be able to render useful production and related advice to farmers as well as carry out farm inspections on request. The workers should always be alert to leverage on innovation which would enhance production processes.

Livestock production clusters may be provided with simple mills to compound feed locally thus reducing costs. The feed-mills could also double as centres for providing animal health-care products and services. Clusters comprised of vegetable and tree-crop farmers could have micro-plants set up to service their primary-processing needs.

Private enterprises on the average are more cost-effective and are run more efficiently than public-sector ventures, especially in the developing countries. All the support-centres are therefore expected to be transferred to private concerns as viable agribusiness ventures, having achieved an acceptable level of profitability and as the new smallholders mature in the science and art of self-sustaining agricultural production.

Marketing

The new smallholders should not initially be compelled to compete with each other or scout extensively for marketing outlets for their produce. Critical to the success of this scheme is the existence of assured markets. Local commodity boards could be established on community bases to evacuate process, weigh and store the produce immediately after harvest. In order to ensure standardization of products and facilitate the grading process, farms within a particular production-cluster should obtain their seed-stock from the same sources.

It would also be the responsibility of the local boards to liaise with regional commodity boards which would have the responsibility of establishing industrial and re-seller markets where buyers can purchase raw or semi-processed produce. The local commodity boards could also organize bazaars and establish consumer outlets where small buyers can routinely take care of their domestic needs.

The use of local boards may prove counterproductive if they are inefficient or set up as profit-centres. Their operational policy should be to operate at near *break-even point*. Unit cost of acquiring the commodity and making it available to buyers should just equal prevailing market price, with the smallholders obtaining the highest possible returns for their efforts. If the local commodity boards are operated purely as profit-centres, there would be tendencies on their part to try to maximize the margin between the final commodity price and their operational costs. They may choose to maximize their returns by keeping acquisition costs as low as possible, thereby discouraging the new smallholders who are yet to be grounded in production. Farmer-groups and cooperative societies should subsequently be encouraged to evolve in order to take over the marketing functions carried out by local commodity boards.

Trading in major agricultural products is a feature of commodity boards and stock exchanges in developed countries. Developing countries should explore this option as a means of expanding their marketing options and increasing farmer-incomes.

Insurance

Compulsory agricultural insurance as well as formal personal insurance policies should be provided for the smallholders at their own expense. This will take care of unfavourable occurrences and ensure the continuity of the scheme by minimizing the adverse effects of hazards which will inevitably occur.

Plans for Disaster-Mitigation should be put in place and the concerned parties should be conversant with them.

Attitudes

Institutions which overtly or covertly shape the political, social, cultural and religious views of the people at the primary and secondary levels should be encouraged to play leading roles in accentuating core-values such as self-reliance, pride in vocational skills and manual labour, love for the earth and respect for its environment and resources. Mass Communications media and Social Media Influencers should also be motivated to come on board.

Attitudinal re-orientation will help reposition small-scale agricultural production to take its rightful place as a respectable profession, facilitate *THE PROCESS* in particular and boost national agricultural production in general.

Chapter Seventeen:

TRANSFORMING SMALLHOLDER AGRICULTURE

Crossroads

Agricultural production in the developing countries of the world, with special emphasis on the emerging economies of sub-Saharan Africa, South America and South-East Asia are still at the crossroads after decades of intervention by their various governments, International Development and other aid agencies. An unforeseen and unintended effect of economic development on the aforementioned economies is that their initial self-sufficiency in respect of agricultural production is giving way to greater dependency on food imports and food-aid in a bid to satisfy their domestic needs.

As new problems emerge, those earlier identified as major constraints to agricultural production in many developing countries decades ago are very much with us today and seem to have assumed even greater dimensions. The inability of the agricultural sectors of these countries to achieve growth comparable with the other sectors of their economies are usually attributed to such causes as willful neglect, civil strife, natural disasters, pests and disease incidence, rural-urban migration, attitudinal issues, urbanization, increasing exploitation of fossil fuels, industrialization and visionless leadership.

The increasing adverse effects of climate change now dictate that governments of especially developing countries must pay greater attention to small-scale agriculture rather than merely treating it as a dumping ground for its under and unemployed citizens.

The governments of many developing countries agree with international development agencies which postulate that an overwhelming majority of the poor people of the world live in rural areas. These economically-disadvantaged people are also predominantly engaged in agriculture-related activities. There also appears to be an ever-widening disparity in incomes and standards of living between the agrarian communities on the one hand and urban commercial centres on the other.

There is a general consensus among the economists of the world that boosting the local economies of developing countries would reduce their general levels of poverty. Paradoxically, the larger proportion of the population of these developing countries appear to be getting poorer with each passing day, even where government revenues have been augmented with the discovery and exploitation of naturally-occurring mineral resources, grants-in-aid and generous cash subvention aimed at reducing the cultivation and sale of psychotropic plants, maintaining historical sites or encouraging wildlife and environmental conservation.

Many developing countries have leisure/tourism and hospitality-based or export-oriented, mono-economies. That is, their economies are to a large extent sustained by tourism-revenues or the export of specific naturally-occurring minerals or produce mostly consisting of non-processed food and fibre. In most of these countries, the typical structure of agricultural production is bimodal. It usually consists of a relatively small, well-supported commercial sub-sector characterized by large mechanized farm holdings, existing alongside a huge, largely-neglected, scattered, rudimentary and near subsistence-level sub-sector, characterized by small fragmented holdings.

It is intended that the awareness of major deficiencies of the agricultural sectors of many developing countries highlighted in this book and the recommended strategies aimed at ameliorating them, will facilitate the transformation of their small-scale farming sub-sector to the advantage of their national economies.

Since every country has its own peculiar circumstances, the ideal approach to agricultural rejuvenation/transformation would be to carefully examine each of the developing countries in order to devise strategies specific to its particular needs. This is however beyond the scope set for this book.

Advantages

The bulk of the food and arguably, industrial raw materials production in most developing countries are carried out by the small farmers who dominate the agricultural landscape of these countries. Indeed, it would be impossible to adequately feed the population of these countries without the small farmer's effective participation in agriculture. Certain reasons can be adduced to show how they have been able to, and have continued to play such an important role.

First of all, methods of traditional small-scale agricultural production are adapted to the local environment and do not require specialized equipment or training. Most rural dwellers are familiar with these methods; hence entry into small-scale production is not restricted to those who have received formal training or tutorials in agriculture, but to a wide spectrum of people of varied backgrounds.

Financing medium to large-scale agriculture is not easy, whereas financial resources required for the initiation and sustenance of small-scale production is within the reach of many individuals. Small-scale production is thus an attractive option for those in search of a new or supplementary occupation. Financing and achieving output expansion is also relatively easier for those engaged in small-scale agricultural production than for many other business ventures.

Small-scale agricultural production activities which are usually carried out in rural communities are able to thrive in the absence of some modern amenities which are the hallmark of economic development. Such amenities include electric power, potable water, telecommunications, all-season road networks and formal banking facilities.

Small-scale agricultural producers possess greater capacity for surviving disasters and unfavourable production circumstances than most large-scale producers. Installed capacity, overheads and other financial obligations being lower for small producers, they can operate at break-even point or can even absorb marginal losses for longer periods of time while they await more favourable circumstances. Small ventures are also easier to reactivate after periods of prolonged inactivity.

Small-scale agriculture is ever ready to absorb the teeming and ever-increasing number of unemployed citizens, especially the able-bodied young to middle-age persons, since it is labour-intensive. Older people are able to engage in small-scale farming until failing health render them physically incapacitated.

In the absence of civil strife, natural disasters and extreme weather and climatic conditions such as drought and flooding, there is usually an array of exploitable opportunities within the agricultural sectors of most of the tropical developing countries.

Agribusiness takes into consideration the sum total of all operations involved in the manufacture and distribution of farm supplies, production operations on the farm, storage, processing and distribution of farm commodities and items made from them. Consequently small-scale agribusiness employs a significant proportion of the population of most developing countries and therefore constitutes integral and important parts of their macro and microeconomic sectors.

The chances of small-scale agricultural production ventures successfully evolving to medium and large-scale, are greater than for most newly established medium to large-scale non-agricultural ventures, since they are usually anchored on firm foundations of experience and expertise. Albeit largely ignored, small-scale agriculture remains a veritable vehicle for national economic growth.

Changes in or adaptations to agricultural production processes are easier carried out on a small-scale. This is because less additional resources are required to facilitate the changes. In addition, the magnitude of the consequences of unfavourable outcomes is lower for small farmers, in direct proportion to the level of their resource-commitment.

Small agricultural producers are more flexible and can easier adapt to innovation and change than medium to large producers because they possess less physical and specialized infrastructure. Once convinced, smallholders are prone to be more adventurous and take greater risks aimed at enhancing their production capabilities, since mistakes may not often prove fatal to the enterprise. Good managerial ability is also able to thrive without being stifled by the structured nature of large-scale enterprise.

Background

It is necessary to review the past and present production circumstances of the typical small-scale farmer in developing countries so as to identify and proffer solutions to surmount the obstacles standing in the way of the rapid transformation of small-scale agricultural production.

With the emergence of international trade and export-oriented industries, many developing countries began to place a lot of emphasis on plantations, ranches and large-scale plants. These are capable of earning significant foreign exchange with which to satisfy acquired tastes for imported luxury items and consumer goods. With liberal credit given by the international community and the exploitation of naturally-occurring mineral resources in some cases, these countries could even afford to import staple food items which they were hitherto self-sufficient in its production.

With the passage of time however, the population of these countries started to increase at a rapid pace while their propensity for consumption also increased. Some of them became heavily indebted to foreign creditors or could no longer balance their national budgets since national expenditures far exceeded national incomes. They have now started to look inwards for ways of reviving and energizing their agricultural sectors in order to satisfy domestic food needs, substitute imported industry raw materials, increase their Gross Domestic Product and increase foreign-exchange earnings.

Operating within the economies of the developing countries is an array of locally and internationally-funded institutions aimed at facilitating the improvement of their agricultural sectors and increasing productivity. These institutions have as their areas of interest; livestock, forestry, fisheries, land resources, planning, cooperatives, cooperative banking, river basin development, agricultural project development, agricultural insurance, seed services, pest control and plant quarantine, stored products research, agro-industrial development, biotechnology and genetic research. The presence of these institutions has not been able over the years to radically transform agriculture in most of these countries. Agricultural transformation obviously necessitates more than the presence of specialized institutions, since there is surfeit or at least adequate numbers of them.

Within most developing countries, there appear to be some dynamic and practical government policies, but these are usually constrained by frequent structural changes and persistent internal bureaucracies. The main problem here appears not to be the absence of well-intentioned plans for the small-scale agricultural sub-sector, but the manner of implementation and lack of continuity. A sizeable number of bureaucrats are quick to assert that the small farmers form the cornerstone of their agricultural transformation programmes, yet this category of farmers is consistently neglected by various government policy implementation agencies.

In many developing countries, relevant authorities and lawmakers deliberately refuse to do the needful because of parochial reasons such as pecuniary self-interest as well as cultural, religious, sectional or political considerations. This tends to slow down the growth of small-scale agriculture even where definite growth paths have been articulated.

An emerging major threat to the small farmer is the growing interest and increasing demand for bio-fuels. The attention of the developed countries is gradually focusing on developing countries as potential sources of *Bio-mass* (bio-fuel feed-stock) which are predominantly staple cereal, legume, shoot and root crops. Ordinarily this should be good news since demand is being expanded. However, vast areas of arable land previously under cultivation by small farmers are gradually being transferred to large-scale producers in the bio-fuels business.

Solutions

As earlier stated, no two developing countries have the same exact problems. Agricultural programmes targeted at small farmers which proved successful in particular developing countries do not necessarily have to be packaged for adoption by other countries, which have their own peculiar production circumstances and are at different stages of agricultural evolution. In each country, agricultural-economists must work with politicians, bureaucrats, engineers, hydrologists, soil-scientists, entomologists, agronomists, sociologists, animal husbandmen, conservationists as well as the small farmers themselves, to devise technical solutions and formulate policies which are economically and politically feasible as well as culturally and socially acceptable. Policy implementation at state and local levels must also be supervised by competent, motivated personnel with adequate background in agriculture.

Low level or under-investment in agriculture, a common feature of tropical developing countries must be addressed. Considerable resources should be employed in order increase or at least maintain current levels of production in the face of increasing natural and man-made calamities faced by the farmer in his production environment. This is also necessary in order to reduce the area of unutilized arable land devoted to agriculture. Responsible land-clearing, land-reclamation and soil restoration should also be carried out after Environmental Impact Assessments. Prospective small-scale farmers should then be encouraged through various incentives to take over the new lands while ongoing efforts to increase the productivity of existing smallholders should be maintained or even intensified.

Agricultural production in developing countries must be stimulated by ensuring that what is produced reach identified markets. Consequently, there has to be increased efforts by the relevant authorities to ensure that the physical infrastructure needed to ensure timely evacuation and facilitate the sale of produce is put in place. These infrastructures include; feeder roads, standard-gauge rail tracks, dredged inland waterways and river ports designed to handle produce which are usually delicate with short shelf-lives and bulky by nature.

Subsidized credit is usually given to medium and large-scale farming concerns with the almost total exclusion of the small farmers from formal agricultural credit in many developing countries. Peasant farmers therefore have to rely exclusively on personal sources of finance or informal credit sources whose interest rates are exorbitant. Small farmers are consequently unable to finance output-enhancing basics such as high-yield varieties, inorganic fertilizers, veterinary drugs, herbicides and pesticides. The over-flogged rhetoric of the risky nature of agricultural production, which fuels the reluctance of formal credit agencies to finance agriculture, must be toned down. Policy measures such as safety-nets in such forms as government guarantee of smallholder loans and promotion of smallholder loan discipline, facilitating the formation of small-farmer thrift and credit associations as well as Agricultural and Personal Insurance, should be formulated and implemented with vigour.

Developing countries should use their increasing numbers of tertiary-education institutions, especially those sited in rural locations, to establish a network of agricultural-vector (disease/pest) and weather-monitoring centres. In addition to random infestations and infections, there are seasonal trends for the upsurge of endemic pests and diseases. Agricultural Biologists and Entomologists should identify and study such *natural rythyms* so that farmers would be better able to predict and handle major incidences.

Local varieties of staple agricultural products resistant to endemic diseases as well as exhibit increased tolerance to extreme weather fluctuation should also be identified by these centres. The gene-pools and data obtained should be used to create templates for the production of genetically-modified or hybrid seed-stock suited to local conditions. Furthermore, efforts should be intensified to find local substitutes for imported raw materials used in food, beverage, leather and agro-allied industries.

There should be conscious efforts to enhance the synergy between small-scale agriculture and the Tourism Industry. During the off-season, some smallholders produce arts and crafts to augument their meager incomes. These objects include cooking utensils, baskets and brooms, pottery, leather and wooden objects, trinkets made of bone and other materials found in the wild which serve as souvenirs for tourists. The tourist potentials of peculiar national agricultural landscapes like those listed as *World Heritage* protected sites should be fully exploited.

Within the developing countries are a sizeable number of local agencies involved in agricultural research. These agencies usually undertake research on improving local varieties, cultivation techniques, farm implements, storage and food processing. The proliferation of these agencies has led to under-achievement, erratic and low level of funding as well as duplication of efforts. Developing countries should evaluate their existing networks of research stations with a view to reducing their numbers and assigning priorities for the surviving units. The surviving units must be well funded and facilities available to them must be upgraded. They should be staffed with qualified, competent, dedicated and well-motivated personnel.

Placing an emphasis on result-oriented agricultural research, there should be an intensification of collaboration and cooperation in the use of available data and resources between local agencies on the one hand and between local and international research agencies on the other.

A significant problem of research is that findings often do not end up in the hands of its intended recipients. Where possible, new devices as well as climate-smart products and processes should be patented and private organizations allowed to mass-produce them for sale to farmers. The research organizations should then work closely with agencies in direct contact with small farmers in order to facilitate the adoption and proper usage of the products / processes.

The conscious or inadvertent entrenched discrimination against the female gender by policy-makers in agriculture should be re-visited. Women are highly significant contributors to the development of the agricultural sectors of developing countries. Given their proven abilities at multi-tasking and very high levels of commitment, they should be given increased access to production resources and greater roles in policy formulation.

Most developing countries have many of their citizens

currently in the *Diaspora*. These are people who emigrated to the developed countries mainly as economic refugees or as students and have established themselves in their adopted countries. Funds remitted by this category of income earners to their families back home amount to a significant proportion of foreign exchange earnings by the developing countries. This largely untapped reservoir of technical expertise and economic power should be exploited in the bid to develop small-scale agriculture. If they perceive a genuine intention by their birth-country's leaders to enhance the fortunes of their people, they would willingly make themselves, their resources, knowledge and influence available. They may very well be the additional ballast that will tilt the balance in favour of tropical small-scale agricultural development.

Developing countries should intensify the use of hybrids and genetically-modified plant and animal species tailored to produce higher yields, have greater resistance to extreme weather, endemic pests and diseases as well as have greater post-harvest longevity. For instance, the cultivation of organic fruits and vegetables do not require the extensive use of pesticides and is more eco-friendly. This reduces production costs and leads to safer and healthier produce. Salt-tolerant varieties of crops would open up greater possibilities for farming communities bordering seas and oceans. There may be trade-offs between Yield and Nutritional Values. Genetic Engineering can help maintain or restore an optimal balance.

Biotechnology must however be approached with caution since the possible health hazards to consumers of genetically-modified food is still the subject of an intense debate. The adoption of genetically-modified species would also leave local farmers largely dependent on the laboratories of the developed countries for their seed-stock. Possibilities of unfair monopolistic pricing tendencies, supply-shortfalls and political considerations exist. The tropical developing countries should therefore collaborate in the establishment of regional functional biotechnology laboratories. These should be given clear mandates, well structured, staffed and equipped to meet the aforementioned and other challenges that may arise.

Developing countries should give greater priority to the slowing-down of the rate of loss of arable land especially through desertification and flooding. Tree-lines and green-belts should be introduced to halt the steady and relentless advance of deserts. In the case of persistent flooding, dredging of water channels should be intensified and smaller dams should be constructed downstream of major dams to control the surge when water is periodically released.

The agricultural landscapes of developing countries are dominated by rural, small and fragmented individual farm holdings. Short-term methods of increasing output revolve around putting additional parcels of land under cultivation through controlled deforestation by fire or mechanized land-clearing, damming and irrigation. Developing countries located in the tropics desiring to increase output by stimulating small-farmer production should de-emphasize the use of heavy machinery, which is even ill-suited to their environment. Greater emphasis should be made on the promotion of simple agricultural machines and tools which are within the means and skills of small farmers. Where it is possible, animal traction could be used to substitute for human muscles and heavy machinery. These measures promote a more efficient utilization of farm energy in small-scale agriculture.

Creating an atmosphere for the acquisition of the right skills, knowledge and attitudes are central to the transformation of subsistence and peasant agriculture in the developing countries. In order to carry out the twin functions of education and manpower-development, agencies responsible must be energized and motivated to produce results. The main targets of their efforts should be the aging urban labour-force, retirees, unemployed and young people willing to make a career in agricultural production. Local languages should ideally be the main or an alternative vehicle for farmer-instruction, while training and retraining programmes for potential and existing small farmers should promote expansion of output without decreasing the productive capacity of the land. This is achieved through rational and efficient use of land and water resources coupled with soil conservation and improvement.

Power available for communications, small transport, processing, storage and preservation in smallholder agriculture should be augmented by the greater use of off-grid clean, renewable energy sources such as solar, wind and water (where practical).

The marked increase in telecommunications awareness and penetration should be exploited to enhance agricultural productivity. Policies inducing telecommunications companies to set up cell-sites in rural locations must put in place. Subsidized or low-cost smart-phones should then be made available to enable small farmers have easy access to production, market and weather-related information. Mobile-money transactions via cell-phones may also be facilitated. Android cell-phones will definitely enhance communication between the farmers, facilitating agencies (credit-institutions, input and service providers, extension-agents, etc) and buyers. Disaster Management activities will also be enhanced. Given the right environment, websites, social-networking sites and phone-in radio programmes where small farmers can freely interact to solve common problems will emerge with time.

Some cultural practices in developing countries serve as hindrances to the development of small-scale agricultural production. Instances abound where exploitable land and water resources are not put to profitable use because they belong to one deity or the other. This includes discrimination against some women who can neither own nor exercise use-rights over agricultural land. Other examples are discriminatory forms of land ownership which leave abundant fertile tracts of lands in the hands of absentee landlords who are not interested in farming. Cultural transformation being a sensitive issue, packages for indigenous communities and small farmers must first conform to their norms and traditions. It should however be designed to gradually wean people from retrogressive beliefs and practices.

Small farmers are continually being displaced to free up land for capital projects such as new towns, dams, airports, petrochemical plants, solid-minerals mines as well as for large-scale agricultural production. These farmers should be relocated to agriculturally viable locations instead of being compensated with paltry sums of money and set adrift. Where the transfers of huge parcels of land to large farming concerns are inevitable, they should be compelled to engage willing displaced smallholders under *Out-grower* (Contract Producer) schemes.

Governments of developing countries should carve out grazing reserves and dedicated routes for migratory livestock. They should also enforce existing laws and enact new ones if needed, mandating the transportation of livestock to markets by rail or trucks to reduce the friction between small farmers and nomadic herdsmen which periodically
lead to loss of lives, properties and disrupted farming schedules. Increased efforts should be made by governments and local authorities to educate farmers on proper methods of cultivation. Inappropriate methods of farming resulting from inadequate knowledge of soil and water conservation especially by small farmers, are gradually exposing large areas of land to wind and water erosion, further depleting and degrading the land available for agriculture.

The daunting task of financing the transformation of small-scale agricultural production in the developing countries is obviously too much for developing countries to undertake unaided. Governments of developing countries should solicit the assistance of international development and other aid agencies to provide counterpart funding for necessary infrastructure such as all-weather access roads and river transport facilities. These agencies could also provide assistance in the provision of primary-processing plants, land-clearing and basic mechanization services such as tractor-leasing. Tackling issues such as; disease/pest control and eradication, soil restoration/detoxification and the control of desertification and flooding typically need the technical expertise and financial resources of the developed economies.

Citing resource inadequacy and expecting sympathy from local and international donors, governments of developing countries pay little or no attention to Disaster Risk Reduction. They should learn to be more proactive than reactive because disasters which disrupt smallholder production activities will occur time after time.

Regional cooperation between communities and countries with common borders is definitely required to disarm, reintegrate and find alternative employment for the significant number of armed militia groups, who regularly go on rampage disrupting rural lifestyles, threatening, robbing and displacing small-scale farmers in some tropical developing countries.

National and local institutions belonging to different political groups but serving the same community should not allow such factors as internal bureaucracy and rivalry to deter them from interacting with each other to improve the lot of small farmers within their sphere of influence. This will prevent the duplication of efforts and ensure the efficient deployment of available resources.

Finally, frequent labour disputes and general insecurity

to lives and physical assets usually caused by leadership ineptitude lead to disruptions in agricultural production, marketing and the supply of critical supplies and services. Establishing equitable political processes aimed at producing competent leaders should help to minimize these incidents which have not been benevolent to small farmers whose products are seasonal, perishable and who lack adequate storage facilities.

CONCLUSION

There is perennial mass-migration of citizens of the tropical developing regions to the developed countries who apart from fleeing from natural disasters and civil strife, may be *economic migrants* escaping from national economies in deep recession or are actually in the search of greater opportunities for self-actualization.

Mass migration unfortunately creates a surplus of intellectual resources in the developed countries while depleting the stock in the developing countries. This deprives the developing countries of high-skill, lower-cost labour needed to attract foreign private-sector investors who help expand their domestic economies.

Mass migration places severe stress on the developed countries arising from problems of economic and socio-cultural integration of the largely uninvited immigrants. The immigrants may also bring with them values at variance with those of their host countries which are potential sources of friction.

Many illegal migrants lack the necessary skills and resources for immediate self-sustenance consequently increasing local crime rates if they are not adequately catered for at the expense of local authourities.

There is dire need to achieve an optimal geographical positioning of world population for greater specialization and efficient utilization of the earth's natural resources. This will also foster better ecological balance and improve global living conditions.

Given increased weather uncertainties due to climate-change, the current food baskets of the world may find themselves unable to meet set targets. It has been observed how armed conflict in one corner of Europe can upset the existing world-wide equilibrium in food and input supplies and consequently prices. It would therefore be prudent to empower tropical developing countries to make up for possible shortfalls.

Despite the formidable natural and human-induced disasters regularly encountered by the tropical developing countries, their populations have been on a steady increase. There is need for these countries to ensure food security for its teeming masses. There is also need for the developed countries to reduce the level of food-aid which they are routinely called upon to dispatch to the tropical developing countries. Though helpful as emergency or short-term relief measures, food-aid does not encourage these countries to be self-reliant.

Energizing small-scale agriculture in the tropical developing regions of the world with its multiplier-effects invariably results in greater food-security, more employment opportunities as well as economic growth which in turn cater for the needs of specialized labour. Sustained and balanced economic growth minimizes the occurrence of civil strife, which is usually fuelled by economic deprivation of the masses and reduces the propensity for mass-migration.

There are huge reservoirs of potential demand for general goods and services in the tropical developing regions of the world waiting to be activated by the economic empowerment of its peoples. Demand is basically the fuel for regional growth. Economic empowerment aimed at stimulating local demand is best achieved in these regions by energizing small-scale agriculture which employs the greater proportion of the population. It hardly comes as a surprise that currently, the faster-growing economies of the world are mostly found in the developing countries. This phenomenon occurs because many developing countries possess vast growth potential.

In the near-saturated domestic markets of the developed countries, re-branding, re-engineering and new regional economic unions in partnership with neighbouring countries have only produced marginal results. Declines in the national *Producer Price Index* (an estimate of the level of prices paid by domestic producers) observed in the developed countries are direct consequences of greater efficiency and stiff competition exacerbating the problem of excess supply over demand. This usually results in deflation which is a precursor to an economic recession. Temporary periods of inflation observed are usually the result of structural imbalances in input and product supply which occur from time to time. Developed countries could find lasting respite by stimulating international demand for their goods and services in the developing countries.

Energizing tropical small-scale agriculture results in a win-win situation for both the developed and developing countries of the world. On the one hand, it creates the right atmosphere for self-sustaining, balanced national economic development and eventual industrialization of the developing countries. On the other hand, new marketing frontiers are made available to the developed countries.

Combined and sustained efforts by the developed and developing countries of the world will provide the needed momentum to energize tropical small-scale agriculture and turn possibilities into opportunities.

REFERENCES

Adebowale K.; *Viability of Traditional Loss Reduction Strategies of Small-scale Farmers:* Business Times, March 21, 1998 !

Adebanjo E.; *Risk Management in Agriculture*: Financial Punch *!

Adesida A; *Quarterly Review* First Bank of Nigeria Ltd.; *

Adeyemo A. A.; *Role of Banks in Agricultural Development*: Business Times. February 29, 1988. !

Adikwu M. A. E.; *Commercial Banks and the Challenges of Agriculture: Business* Concord. January 1, 1988. !

Ahamefula C. H.; *Good Storage Practice-Key to Good Quality Feed Production:* Livestock Farmer. Volume 14, No 3 September 1989.

Akpabio T.; *Agricultural Credit Guarantee Scheme:* Business Times *,!

Amaefule Everest.; *Formal Agricultural Credit: Sunday Punch* December 14, 2008. !

Antonios P.; *Agricultural Development and Environmental Protection -The Challenge to Africa:* Business Concord October 17, 1989. !

Ayodele F. and P. N Vine; *Elements of Soil Science 11*: Department of Agronomy, University of Ibadan*

Bank of Agriculture Website boanig.com Accessed: September, 2022

Bello T. M.; *Towards Effective Management of Agricultural Loans*: Business Times. September 8, 1986. !

Boco F.; *Effect of Storage Environment on Food Products*: Business Concord. March 31, 1987. !

Boco F.; *Food Production Problems in Nigeria*: Business Concord. *,!

Bokime B.; *Agricultural Credit: The A.C.G.S.F. Profile*: Guardian Financial Weekly May 2, 1988. !

Bolaji W.; *Food-A Twin of National Development*: Business Concord. September 26, 1989. !

Bureau of Public Enterprises Website *who we are* www.bpe.gov.ng /nigerian-agricultural-cooperative- and- rural-development-bank/nacrdb Accessed: September, 2022

Celia A. Harvey et al; *Extreme Vulnerability of Smallholder Farmers to Agricultural Risks and Climate Change in Madagascar* http://:rsth.royalsocietypublishing.org. 20130089.full#content-block. Accessed: May, 2014.

Central Bank of Nigeria Website; http://: www.cenbank.org. May 2, 2006. Accessed: June, 2012.

Cobley L. S. and W. M. Steele; *An Introduction to the Botany of Tropical Plants*: Second Edition. Longman 1976

Dedekuma E.; *Financing Agricultural Cooperatives (Compilation)* Business Concord. October 4, 1988. !

Dedekuma E.; *How to Make Cooperative Societies Work (Compilation): Business* Concord. October 4, 1988. !

Davis J. and R. Goldberg.; *A Concept of Agribusiness: Boston, Mass.* Research Division, Harvard Business School. 1957

Edema A.; *Risk Management in Agriculture*: Financial Punch. *, !

Ekula E.; *Corporate Risk Management*: Business Concord. June 13, 1989. !

Ezekiel E.; *Achieving Self-Sufficiency in Food Production*: Sunday Punch July 3, 2011. !

Fakorede L.; *A Small Help to Small-Scale Businesses*: Business Concord.*,!

Foyo Y.; *Of Nigerian Banks and Nigeria's Agriculture*: Business Concord. *, !

Horoszowski M.; *The Empowerment Line - A new measure to Fight poverty (and move beyond it).* February 21, 2014 blog.movingworlds.com. Accessed: April 2015

Idachaba F. S; *State-Federal Relations in Nigeria Managing Agricultural Development in Africa* Madia Discussion Paper 8, USAID World Bank, June 1989

Idowu O. O.; *Agricultural Insurance Principles and Practice:* Business Times. April 18, 1988. !

Ifedi C.; *Business Failures: Business* Concord, November 4, 1986 !

International Fund for Agricultural Development; *Adaptations for SmallHolder Agricultural Programme:* www.ifad.org.asap.pdf. *Accessed*: May, 2014.

International Institute of Tropical Agriculture; *Live Mulch: Long-Term Effects on Crop Yields, Weed Infestation and Soil Organic Matter.* Annual Report and Research Highlights, ISSN 0334-4340. Ibadan, Nigeria. 1985.

International Institute of Tropical Agriculture; *Evaluating Savannah Soil*: Annual Report and Research Highlights, ISSN 0331-4340. Ibadan, Nigeria. 1984

International Institute for Sustainable Development; *Climate Risk Management for Small-Holder Agriculture in Honduras* http://:iisd.org honduras.pdf. *Accessed*: May, 2014.

Keshi L.; *Why Our Businesses Fail*: Business Concord. March 27, 1987,!

Kohls R. L.; *Marketing of Agricultural Products: Macmillan* 1968.

Kollere M. A.; *Role of Financing Institutions*: Business Concord, October 13, 1987. !

Lanzi A. I ; *Marketing Scene-Towards Getting Your Target Market:* Business Times. December 21, 1987. !

Lauf P.; Agricultural Geography. Vol. 1. *Systems, Subsistence and Plantation Agriculture*: Nelson. 1968.

Miller L. F.; *Agricultural Credit and Finance in Africa:* USA. Rockefeller Foundation. 1977.

Ministry of Agriculture (Information Unit); *Anambra State in Food Production: Daily* Star. April 15, 1986. !

Morgan W. B.; *Agriculture in the Third World-A Spatial Analysis*: Bell and Hyman. London. 1980.

Mosher A. T.; *Getting Agriculture Moving-Essentials for Development and Modernization:* New York. Published for The Agricultural Development Council by Fredrick A Praegar. 1966.

Myint H.; *Economics of Developing Countries*: (5th Edition) London. Hutchinson. 1980

Nigerian Agricultural Insurance Corporation. Website: www.naic.gov.ng. Accessed: September, 2022

Nigerian Agricultural Quarantine Service (NAQS) Website www.naqs.gov.ng Accessed: September, 2022

Nigerian Incentive-Based Risk Sharing System for Agricultural Lending Website nirsal.com/who we are Accessed:: September, 2022

Nwalieji H.U. & Igbokwe E. M; *Role of Local Governments in Agricultural Development in Nigeria - A Review* Journal of Agricultural Extension Vol. 15(2) December, 2011

Olatubi W. O.; *Setting Research Priorities in Nigeria's Agriculture: Business* Times. October 26, 1987!

Olatubi W. O.; *Frontiers in Nigeria's Agriculture: Business* Times. September 1, 1986. !

Oni O.; *Risk Management in Agricultural Investments: Business* Times. February 26, 1989. !

Onuba Ifeanyi; *CBN's Re-Engineering & Financial Literacy of the Grassroots:* Sunday Punch. February 7, 2010. !

Onuba Ifeanyi; *Taking Agriculture to the Threshold of Growth:* Sunday Punch July 17, 2011. !

Onyemachi S. U.; *The Small Farmer and the Need for Cooperation:* Business Times. September 29, 1986. !

Owuamanam Jude; *CBN Disburses ₦134Bn to Farmers*: Sunday Punch July 17, 2011. !

Patterson R. F.; *The Varsity English Dictionary: Vasity Publications. London.* *

Phillips J.; *The Development of Agriculture and Forestry in the Tropics- Patterns, Problems and Promise.* Faber and Faber. 1961.

Rome Partnership for Disaster Risk Management; *Disaster Risk Management in Food and Agriculture: http://:home.wfp 201794.pdf Accessed:* May, 2014.

Nigerian Stored Products Research Institute; *Storing your Produce Advisory Booklet No. 1, Maize* (1982) *Advisory Booklet No. 2, Yam Tubers and Dried Yam* (1982) *Advisory Booklet No. 3, Cassava and Garri* (Revised 1983)

Shaka B.; *Beyond the Limits of Agricultural Insurance*: Business Times. November 21, 1988. !

Tel D. A. and Hagarty M; *Soil and Plant Analysis Study Guide for Agricultural Laboratory Directors and Technologists working in Tropical Regions* International Institute of Tropical Agriculture and University of Guelph; 1984.

UCD Library; *MLA Referencing Style: www.wcd.ie/library.* Created: 25th July, 2011. Dublin. Accessed: July 2, 2014.

Ugheoke E.; *Boosting Agricultural Productivity in Nigeria, Policy or Practice: Financial* Punch. July 2, 1986. !

Umokoro F. E.; *Financing the Small Farmer through Effective Seasonal Credit: Business Concord.* October 13, 1987.!

Umokoro F. E.; *Avoiding Risk and Uncertainty in Agriculture Business Concord.* July 21, 1987. !

Umokoro F. E.; *Increasing Farm Labour Productivity through Better Use of Energy: Business Concord.* May 1, 1987!

United Nations Office for Disaster Risk Reduction; *(The) Disaster Risk Reduction:* http://:www.unidr.org *Accessed*: May, 2014

Wikitionary; http://:en.m.wikipedia.org *Accessed*: May, 2015.

* Date Uncertain.   ! Page Uncertain.

INDEX

ACGSF 40
ADPs 39
Agribusiness 158
Agricultural Finance Intermediation 41
Agricultural Insurance
   -definition 53
   -forms 53,54
   -indemnity 55
   -premium 54,55
Agricultural Value Chain 35,96,141
BOA 42
Bio-mass 161
Break-even Point 78,153
Brokers 139
Cash Cropping 13
Cash Sufficiency 86
CBN 39
Competition Factor 45
Consumer-Price Index 145
Contract-Farming 44,158
Cooperation 22,90
Costs
   -control 86
   -fixed 73
   -variable 73
Credit
   -consumption 26
   -cost 30
   -group 21
   -informal 26
   -production 16,25
   -sources 26
Crib 127
Diversification 50
Economic Migrants 171
Extension
   -agents/workers 100,104
   -philosophy 99,103,
   -role 100

Farm Busines 14
Farmer Multipurpose Cooperatives 21
Farming Technology 18
Fertilizers 116
Finance/Funding Sources 69
Fixed Costs 73
Flushing 117
Grain Store 133
Group Action
   -condition 89
   -forms 90
Growth Media 123
Hazards 59
Ideal Crop 118
Inter-crop 121
Intermediation 56
Irrigation 117
Item Definition 72
Leaching 118
Liming 120
Loss 123,125
Marketing
   -definition 136
   -functions 141
   -Margin 138
MGI Empowerment Line 12
Micro-finance 40
Middlemen
   -facilitating 139
   -merchant 138
Moisture Content 131
Mulching 119
NACRBD 42
NAIC 41
NAQS 42
National Agricultural Sector Objectives 37
NIRSAL 42
NPFS 42

Organic Matter 116
Peasant Farmer 13, 91
Per Capita Income 11
Premium 54
Planning 67
Post-Harvest 123
Produce Processing 140
Producer Price Index 173
Production Medium 123
Profit Motivation 82
Puddling 119
Quantitative Analytical Tests 79
RBDA's 39
Risk
  -analysis 54
  -control 54
  -definition 44, 53
  -identification 53
  -management 47, 53, 86
  -planning 49
Rochdale Principles 22, 97
Seasonality 136,142
Sensitivity Tests 78, 84
Silo 130
Small farmer/Small-scale/Smallholder 14
Social Acceptability 70

Soil
  -abuse 113
  -classification 111
  -conservation 114
  -definition 111
  -functions 111
  -maintenance 117
  -management 115
Subsistence 12
System Definition 72
Stagflation 145
Strip Farming 117
Technical Feasibility 69
Thrift and Credit Unions 21
Time Value 74
Uncertainty
  -definition 44, 47
  -institutional 48
  -legislation 48
  -planning 49
  -price 48
  -yeild 48
Variable Costs 73
Wants 16
Weather Factor 44

www.ingramcontent.com/pod-product-compliance
Lightning Source LLC
Chambersburg PA
CBHW052350220526
45465CB00003BA/1035